Proceedings

FH Science Day

Proceedings

FH Science Day

25th October 2006

Workshops: How Emotions Influence Decisions and Behavior

Technical Applications in Medicine

Logistics and Supply Chain Management

Operations Management

Secure and Embedded Systems

Editors: Prof. (FH) DI Dr. Jürgen Ecker

Prof. (FH) DI Dr. Herbert Jodlbauer

Prof. (FH) DI Dr. Johann Kastner

Prof. (FH) Dr. Jörg Kraigher-Krainer

Prof. (FH) DI Dr. Thomas Müller-Wipperfürth

Prof. (FH) DI Franz Staberhofer

Prof. (FH) DI Dr. Martin Zauner, M.Sc.

Shaker Verlag

Aachen 2006

Bibliographic information published by Die Deutsche Bibliothek
Die Deutsche Bibliothek lists this publication in the Deutsche
Nationalbibliografie; detailed bibliographic data is available in
the internet at http://dnd.ddb.de.

contract address:

FH OÖ Forschungs & Entwicklungs GmbH
Franz-Fritsch-Straße 11/Top 3
Tel.: +43 (0)7242/44808-40
Fax: +43 (0)7242/44808-77
E-Mail: research@fh-ooe.at
www.fh-ooe.at

Layout: MMag. Monika Dopler

Copyright Shaker Verlag 2006
All rights reserved. No part of this publication may be reproduced, stored in a
retrieval system, or transmitted, in any form or by any means, electronic,
mechanical, photocopying, recording or otherwise, without the prior
permission of the publishers.

Printed in Germany.

ISBN-10: 3-8322-5555-9
ISBN-13: 978-3-8322-5555-8
ISSN 1432-4385

Shaker Verlag GmbH • P.O. BOX 101818 • D-52018 Aachen
Phone: 0049/2407/9596-0 • Telefax: 0049/2407/9596-9
Internet: www.shaker.de • eMail: info@shaker.de

Introduction

The Upper Austria University of Applied Sciences was established in the year 1994. Since that time there are placed about 29 several academic study programs at four locations in Upper Austria. The Upper Austria University of Applied Sciences is an academic training faculty with high research potential.

The Upper Austria University of Applied Sciences (abbr.: FH OÖ) has reinforced its research and development activities in recent years by establishing a separate enterprise and four R&D-Centres at the FH OÖ locations Hagenberg, Linz, Steyr and Wels.

The main purpose of the Upper Austria University of Applied Sciences Research and Development Ltd (FH OÖ Forschungs & Entwicklungs GmbH) is to enhance the competitiveness of the co-operation partners through innovation and new technologies, and to support the business location Upper Austria through the implementation of research results.

At the FH Science Day 2006 the latest project results and scientific discussions with international representatives take centre stage.

This year the FH Science Day focuses on the following subjects:

- How Emotions Influence Decisions and Behavior
- Technical Applications in Medicine
- Logistics and Supply Chain Management
- Operations Management
- Secure and Embedded Systems

Dr. Gerald Reisinger
CEO

Prof. (FH) DI Dr. Johann Kastner
Management

Contents

How Emotions Influence Decisions and Behavior.................... 1

Jörg Kraigher-Krainer
A Framework for Conceptualizing Purchase Emotions.................... 3

Hans-Peter Liebmann, Jörg Kraigher-Krainer
Can the Emotion Construct Help Understanding of Contemporary Phenomena in Knowledge Management?...................... 22

Claudio Boido, Antonio Fasano
Football and Mood in Italian Stock Exchange............................. 35

Technical Applications in Medicine.. 51

Sirid Griebenow, Gebhard Rieger, Jutta Horwath-Winter, Otto Schmut
A New Method to Determine the Sustainability of the Increased Watersoluble Antioxidative Status (ACW) in Tear Fluid Obtained without Stimulation after Iodide-treatments in Bad Hall........................ 53

Alexander Müller, Ulrike Fröber, Stefan Lutherdt, Hartmut Witte
Concepts of Biomechatronics: Individualized Prevention of Hearing Loss in the Working Environment and in Daily Life....................... 57

Reinhard Hainisch, Stefan Froschauer, Harald Schöffl, Martin Zauner
Application of a Miniature Endoscope to Investigate the Inner Surface of Blood Vessels... 60

Christoph Guger, Robert Leeb, Doron Friedman, Vinoba Vinayagamoorthy, Angus Antely, Günter Edlinger, Mel Slater
Controlling Virtual Environments by Thoughts............................. 66

Werner Backfrieder, Wilma Maschek
Image Registration of MR/CT and Low-Count SPECT.................... 69

Franz Pfeifer, Roland Swoboda, Gerald Zwettler, Werner Backfrieder
RAPS: A Rapid Prototyping Framework Based on RCP, ITK and VTK... 74

Logistics and Supply Chain Management.................................. 81

Sabine Bäck, Uwe Brunner
How Can Individual Procurement Processes Fit in Comparable Supply Chains?... 83

Klaus Schmitz
MCSP – Multi Customer Supplier Park
A Solution Approach for the Automotive Supply Industry in CENTROPE/AREE.. 89

Péter Németh, József Marek
RFId Implementation in a Logistics Distribution Center.................... 95

Wolfgang Greisberger, Corinna Engelhardt-Nowitzki
Organizational Learning as a Dynamic Driver for the Development of Logistics.. 100

Tatjana Wallner, Alexander Stüger
Supply Chain Strategy and Organization... 107

Corinna Engelhardt-Nowitzki, Helmut E. Zsifkovits, Elisabeth Lackner
Critical Success Factors for Supply Chain Collaboration and Supply Chain Governance in Complex and Dynamic Environments: A "Design for Agility".. 115

Evelyn Rohrhofer, Karin Spaeth
The Steyr Network Model as a New Management Approach for Modern Logistics... 127

Operations Management.. 137

Klaus Altendorfer, Herbert Jodlbauer
The Concept of „Customer Driven Production Planning" for Evaluation of Job-Shop System Performance.. 139

Wolfgang Promberger, Herbert Jodlbauer
Makespan Minimization for Parallel Machines................................. 150

Andreas Weidenhiller, Gabriel Kronberger
New Developments in Scheduling - Improved Model and Solver Approaches... 153

Gabriel Kronberger, Andreas Weidenhiller
Automated Generation of Simulation Models from ERP Data – An Example.. 160

Martin Schöffer, Christian Weger
Total Effective Equipment Productivity – The Key to Maximizing the Efficiency of the Industrial Production... 166

Secure and Embedded Systems... 173

Oliver Hable, Barbara Hauer
NetScanAssistant – Platform Independent Network Scanner as Plug-In in Eclipse... 175

Johann Wallinger
Austrian Mobile Equipment Tracking System (AMETS).................. 180

Gerald Madlmayr, Oliver Dillinger, Sebastian Gierlinger, Peter Kleebauer, Roland Pucher, Dieter Vymazahl, Josef Langer, Christoph Schaffer, Jürgen Ecker
Secure Token Concept within a Near Field Communication Ecosystem... 188

Martina Heiligenbrunner, Daniel Slamanig, Christian Stingl
Analysis of Austrian Citizen Cards.. 194

Andreas Gaupmann
Personal Firewalls for Linux Desktops...................................... 204

Oliver Dillinger, Gerald Madlmayr, Christoph Schaffer, Josef Langer
An Approach to NFC's Mode Switch.. 210

Werner Kurschl, Stefan Mitsch, Rene Prokop, Johannes Schönböck
Model-Driven Development of Speech-Enabled Applications... 216

Thomas Plachy
Evaluation of Temperature-aware Quality of Service..................... 224

Index of Authors... A

Workshop:

How Emotions Influence Decisions and Behavior

Chairman: Jörg Kraigher-Krainer

Scientific Board:
Jörg Kraigher-Krainer, FH OÖ Campus Steyr
Hans Peter Liebmann, Universität Graz
Gerhard Wührer, Universität Linz

A Framework for Conceptualizing Purchase Emotions

Jörg Kraigher-Krainer [a]

[a] Upper Austria University of Applied Sciences, Campus Steyr, Wehrgrabengasse 1-3, A-4400 Steyr, AUSTRIA

ABSTRACT

Motivation: In recent years, considerable effort has been devoted to better understanding the impact of emotion on cognition and decision making. Yet, comparably little is known about the antecedent and subsequent conditions of emotions during the purchase process.

Results: A literature review, a definition of the term *purchase emotion*, a conceptual framework and a set of hypotheses are offered. It is proposed to understand emotion and motivation as an emerging psychic process starting with a schema discrepancy and developing to intrinsic vs. extrinsic motivation, and, together with cognitive processes, to different emotional qualities. The presented conceptualization could help to (1) better integrate several theoretical and empirical contributions within the field of emotion psychology and beyond it, (2) stimulate future research about the impact of emotion on consumer behavior and (3) help practitioners to better understand their core business.

Contact: joerg.kraigher@fh-steyr.at

1 INTRODUCTION

The dominant picture in consumer behavioral research still is that people either like the product they are aiming for – e.g. high involvement products – or at least have some indifferent attitude toward that product – e.g. low involvement products. Furthermore, it is frequently assumed that innovation, assortment, or information, are unconditionally welcome and satisfying activities, resulting at first in watchful satisfaction and finally in increased revenues or profits. Anyway, recent attempts in literature show, that consumer decisions can be accompanied by negative emotions resulting from mental inconvenience, information overload, or decision complexity, especially if the product or service is perceived as not being worth the time for extensive investigation.

The research question in this article is: Under which conditions can we expect consumers to experience emotions during a purchase decision process and how could these emotions influence the process? Chapter two gives a brief literature review on the problem at hand and defines the term *purchase emotion*. In Chapter three the emotion construct, its antecedents and its consequences are conceptualized. Chapter four presents a set of hypotheses and gives reasons for them, whereas in chapter five the managerial implications are discussed.

2 SYNOPSIS OF RELATED LITERATURE ABOUT PURCHASE EMOTIONS

Emotions are a matter of systematic behavioral research – at best – for the last two or three decades (Zajonc 1980). Anyway, in recent years, considerable effort has been devoted to understanding, how emotions affect decision processes (see e.g. Andrade 2005; Bagozzi, Gopinath and Nyer 1999; Laros and Steenkamp 2005 for contemporary reviews). These decades of research have produced a rich body of empirical data as well as models. Despite these contributions, important theoretical, practical, and methodological questions have not been addressed, especially on the negative side of emotions.

Domain of the problem. From a *functional* perspective (see Laros and Steenkamp 2005, Nyer 1997) emotion research is discussed in contexts like Advertising (e.g., Brown, Homer and Inman 1998; Gorn, Pham and Sin 2001); Customer Relationship Management (e.g., Reichheld 1997, 2004; Nyer and Gopinath 2004); Corporate Social Responsibility (e.g., Loewenstein 1996; Tice, Bratslavsky and Baumeister 2001); or POS-Management (e.g., Adaval 2001; East et al. 1994). From a *research* perspective it is rather difficult to clearly define a domain of the emotion construct, as more than one of the prominent constructs in consumer behavior claim for an emotional component contrasted to a cognitive one, for instance Involvement Research (e.g., Mittal 1987); Perceived Risk Research (e.g., Dowling and Staelin 1994) or Attitude Research (e.g., Batra and Ahtola 1990). Taking into account that the number of papers concerning emotional phenomena is now dramatically increasing it seems recommendable to understand emotion research as a young and already independent field of research, hence as the domain of the research attempt lying before and dealing with constructs like *mood*, *feeling*, *emotion*, *affect*, *motivation* and related terms.

Defining the term *purchase emotion*. Emotions (a) give rise to affective experiences, especially feelings of arousal and pleasure/displeasure; (b) generate cognitive processes such as emotionally relevant perceptual effect, appraisals, labeling processes; (c) lead to behavior that is often, but not always, expressive, goal-directed, and adaptive; and (d) have a communication function within the social environment (for similar definitions see Kleinginna and Kleinginna 1981, p. 355; Frijda, Manstead and Bem, 2000, p. 5). *Purchase emotions* are distinguished from the term, *consumption emotions,* insofar as the latter term is sometimes understood as emotions during the consumption process (Chaudhuri 1997) or as emotions rising from consideration, buying, and using products and services (Cohen and Areni 1991; Ruth, Brunel and Otnes 2002), whereas *purchase emotions* focus on emotions arising during the pre-purchase and decision process of buyers, rather than emotions during the consumption process or the post purchase-phase, respectively. Concentrating on the pre-purchase and decision process of *buyers* also means that this definition focuses on buyers and not on everybody – as for instance involvement research or attitude research do.

Although the discussion about the differences or similarities between terms like *emotion, affect, feeling*, or, *mood* is far from being resolved, in this context I only distinguish *emotion* from *mood*, where I talk of mood, if an emotion is (1) enduring (hours to weeks), (2) not related to a specific entity, and (3) mainly subconscious or of peripheral consciousness (see e.g. Bagozzi, Gopinath and Nyer, 1999; Gendolla 2000; Gross 1998; Pham 1998; Schwarz and Clore 1983; 2003). Mood has been proposed to serve similar to background information contrasted to the figural impact of emotion (Hänze 1998, Pham 1998), e.g. elicited by a certain product. Hence, it can also be assumed, that emotions have more impact on judgment than mood (Schwarz and Clore 2003) and that the probability of misattributing extraneous feelings to an object of judgment is higher for mood, than for emotion.

Hedonic man assumption. It is theoretically and empirically well supported that man seeks pleasure and avoids pain (Andrade 2005; Gendolla 2000; Holbrook and Hirschman 1982; a recommendable contemporary review is offered by Tice, Bratslavski and Baumeister 2001). Isen and her colleagues have introduced the term *mood management* consisting of two activities: (1) *mood maintenance* to adhering to positive mood and (2) *mood repair* to getting over negative mood (see e.g., Isen and Simmonds 1978; Isen and Geva 1987; for a review concerning mood management see Gross 1998).

Emotion and motivation. There seems to be a close relationship between the terms *emotion* and *motivation* (Gendolla 2000) on the one hand and between the terms *motivation* and *mood management* on the other hand: if we pursue promotion goals, we seek pleasure and if we pursue prevention goals we aim to avoid pain (Aaker and Lee 2001). Always seeking pleasure following ones impulses can cause future pain. Therefore our decisions have to be balanced between momentary impulses and future oriented impulse control (Tice, Bratslavsky and Baumeister 2001). Following Ryan and Deci (2000, p. 55) the former motivation is called *intrinsic motivation*, as it refers to doing something because it is inherently interesting or enjoyable, while the latter is termed *extrinsic motivation*, which refers to doing something because it leads to a separable outcome. For attempts to transfer Deci and Ryan's (1985) Self-Determination-Theory into the marketing sector see for instance Babin, Darden and Griffin (1994), Holbrook and Gardner (1998), Van Trijp, Hoyer and Inman (1996). Although some authors take the emotion-motivation-relationship for unresolved (e.g., Schwarz and Clore 2003), an increasing community equates the two terms (e.g., Adaval 2001; Bosmans and Baumgartner 2005; Fiedler and Bless 2000; Gendolla 2000), understanding emotions as basic, primal motivations (Chaudhuri 1997, p. 82), a position which corresponds with that taken in this paper.

Cognition and emotion. There is a – rather academic – discussion about the question, whether emotions are triggered by (*appraisal theories*, e.g., Bagozzi, Gopinath and Nyer 1999, Lazarus 1991, Nyer 1997) or cause cognitions (*affective primacy hypothesis*, e.g., Forgas 1995; Schwarz and Clore 1983; Zajonc 1980). Why do I call this discussion *academic*? Nyer (1997), for instance, asks for the antecedents of emotions (which are cognitions in his opinion). But this raises the subsequent question: what are the antecedents of these antecedents? A likely solution to this vicious circle is the replacement of the S-O-R-logic, which apparently underlies this discourse, by a cybernetic loop-paradigm, where emotions elicit *and* arise from appraisals (for loop-models see Bower 1981, Isen et al. 1978; Kraigher-Krainer 2005; Lerner and Keltner 2000).

Construct dimensionality. Most classifications of emotions provide two dimensions, most of those, again, provide a positive-negative-dimension (Bagozzi, Gopinath and Nyer, 1999; Hänze 1998). Although some authors scrutinize the reduction of emotions to its positive-negative-dimension (e.g., Holbrook 1986), it appears to be the most popular conceptualization (Laros and Steenkamp 2005, p. 1439 f). Furthermore, this approach is in correspondence with findings in brain research (e.g., Morse 2006; Olds and Milner 1954) and with our everyday experience. Another point of discussion, which is still unclear, regards the question, whether the positive-negative-dimension itself is bipolar or bidimensional. For instance, Brown, Homer, and Inman (1998), in their review, discuss the bipolarity hypothesis contrasted to the bidimensional hypothesis, and rather find support for the latter. Bagozzi, Gopinath and Nyer (1999) conclude that it is still unclear whether pleasant vs. unpleasant are two independent dimensions or, the two poles of one dimension. But they commit that everyday experience would favor the bipolarity hypothesis, as one is either happy or sad. For the purpose of this paper the construct is understood as bipolar, reaching from a positive/pleasant to a negative/unpleasant pole.

3 CONCEPTUALIZING PURCHASE EMOTIONS

Initial Emotion. Any motivational process has to start somewhere, as we are not permanently motivated. Yeung and Wyer (2004) talk of an initial impression and Lerner and Keltner (2000), in their *appraisal tendency theory*, of an incidental affect. For the conceptualization lying before it is proposed to think of an *initial primal emotion*, which is triggered by an expectation discrepancy (Heider 1958; Matzler 1997; Oliver

1980); or a schema discrepancy (Derbaix and Vanhamme 2003; Glasersfeld 1997; Powers 1973) between a subjects private theory or expectation and his/her perceptions. This initial primal emotion is best described as *surprise* (Derbaix and Vanhamme 2003) and has two qualities: positive, pleasurable surprise, if the perception exceeds the expectation, and negative disappointing surprise, if the perception is less than expected.

Surprise is understood here as a short-lived emotion, that leads to physiological (arousal, skin conduction) and psychological (attention, curiosity, cognitive work) reactions (Derbaix and Vanhamme 2003). It has (1) an interrupting function on whatever we are just doing (Damasio and Damasio 1994; LeDoux 1996), (2) an evaluative function as it signals potential personal harm or benefit (Ciompi 1998; Lazarus 1991; Zajonc 1980), (3) a motivational function as it provides the organism with energy to think and act (Lazarus 1991, Gendolla 2000), and, (4) an adaptive function to update our schemas (Derbaix and Vanhamme 2003; DeSteno et al. 2004; Fiedler and Bless 2000; Glasersfeld 1997; Gross 1998; Lazarus 1991). These emotions are not *transferred* to an informational system, where they are judged as good or bad, positive or negative. They *are* information themselves (*preferences need no inference*, Zajonc 1980) and represent the good-bad-dimension of our psychic system.

As long as experiences meet our schemata, the organism is in the state of habit behavior, driven by established routines and acting the way it acted in the past (see Fig. 1: No schema discrepancy → habitual behavior). Anyway, if a primary emotion of surprise signals environment-schema-deviations, the attention is drawn to this deviation and energy is mobilized to investigate further. Therefore, it is assumed, that there is no cognitive processing without such an initial emotion. Anyway, there is an ongoing mutual influence between emotions and cognitions as soon as this initial emotion has elicited a response from our conscious psychic system.

Subsequent Cognitions and Emotions. The cognitive system is here understood as a system that controls human resources and potential loss of these resources with its most radical potential loss: the loss of life. This is why a cognitive system requires self consciousness – for all we know, man is the only animal being aware of the fact, that it can and will one day lose its life. Fortunately, there is a *core concept for consumer theory* (Conchar et al. 2004, p. 418) that deals with potential loss of consumers: the perceived risk concept (Bauer 1960).

As the focus of this paper is on the emotional side of decision making, the cognitive aspect is not further explicated. But it is important to note that there have been attempts to find a systematic relationship between emotional experiences and risk perception (e.g., Isen and Geva 1987) leading to the consistent and contradictory result that positive mood leads to more optimistic and more cautious action tendencies at the same time (Nygren et al. 1996). As Isen and Geva (1987) have pointed out, the minimum relationship between these constructs rests on the fact, that if one is in a good mood, he inevitably has one thing to lose, namely this good mood (see mood maintenance hypothesis above). But as emotion has been separated from mood (see above) it can nevertheless be stated, that under an emotion-focused view of the construct it is better to grasp perceived risk as being basically independent and distinct from the emotional one, like Lerner and Keltner (2000) or Gendolla (2000) propose.

Anyway, ongoing emotion-cognition-loops seem to be the source of the different emotional qualities we experience in everyday life. An initial emotion of negative, disappointing valence combined with perception of high risk might lead to the experience of fear, whereas combined with low risk might elicit anger (Lerner, Small and Loewenstein 2004). Similarly, it can be assumed that an initial emotion of negative, disappointing valence will cause attribution processes, considering aspects like task difficulty, effort, or other persons contributions and subsequently lead to differentiated emotions like guilt (Ruth, Brunel and Otnes 2002); regret (Herrmann,

Huber and Seilheimer 2003; Zeelenberg and Pieters 2004); vent (Nyer and Gopinath 2005); or, disgust vs. sadness (Lerner, Small and Loewenstein 2004). And, as from enduring positive emotion-chains intrinsic motivation emerges, whereas from enduring negative emotion-chains extrinsic motivation emerges, it can also be assumed that under intrinsic vs. extrinsic motivation emotional qualities like fun vs. boring, useful vs. useless, cheerfulness vs. dejection, or quiescence vs. agitation provide different informative validity for the organism (Bosmans and Baumgartner 2005; Pham 1998).

Although I have proposed to replace the S-O-R-paradigm by a cybernetic-loop-model, to gain a better understanding of the way, emotions during purchase decision processes influence cognitions and processing styles in a dynamic way, the framework depicted in Fig. 1 separates antecedents from consequences of emotions. This is primarily for didactical reasons, being aware of the fact that loops actually have no starting or ending point and fully agreeing with authors like Scherer (1995) who see emotions as an interdependent, interrelated flow of consciousness with sudden changes and high plasticity, hence overstretching any attempt to freeze them by cause-effect-diagrams.

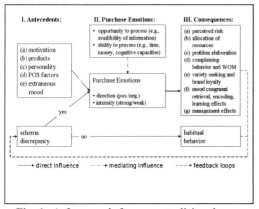

Fig. 1. A framework for conceptualizing the purchase emotion construct.

I. Antecedents of purchase emotions. There are two antecedent conditions for the experience of an emotion during a purchase decision, depicted in Fig. 1: (1) schema discrepancy and (2) in case there is a schema discrepancy, a set of antecedents consisting of (a) motivation; (b) product and product class; (c) personality; (d) POS factors; and (e) extraneous mood, which are assumed to have a direct impact on the experience of purchase emotions.

(a) Motivation. The influence of shopping motives on emotions is theoretically and empirically well supported. Although under different terms like hedonic vs. utilitarian (Babin, Darden and Griffin 1994; Batra and Ahtola 1990; Holbrook 1986; Holbrook and Gardner 1998; Mittal and Lee 1989), entertaining vs. nonentertaining (Jones 1999), fun side vs. dark side of shopping (Sherry, McGrath and Levy 1993), consummatory vs. instrumental (Batra and Ahtola 1990; Holbrook 1986; Holbrook and Gardner 1998, Pham 1998), hedonic vs. rationalistic (Trommsdorff 1993), achievement vs. avoidance (Bosmans and Baumgartner 2005), approach vs. protection (Bosmans and Baumgartner 2005), appetitive vs. aversive (Bosmans and Baumgartner 2005), promotion vs. prevention (Aaker and Lee 2001), they all seem to tap the phenomenon which has earlier been described by the basic idea of intrinsic vs. extrinsic motivation (Deci and Ryan 1985; Ryan and Deci 2000). It has also been mentioned that motives, strictly speaking, cannot be separated from emotions as they are the product of ongoing emotion chains (see feedback loops in Fig. 1).

(b) Product and product class. Closely related to the motivation is the question which product or service has to be purchased. Accordingly, Wells (1986) speaks of avoidance products and approach products with utilitarian products being located somewhere in between the two other categories. Similarly, Yeung and Wyer (2004) and Schmalen (1994) contrast hedonic to utilitarian products, or Rossiter, Percy and Donovan (1991) distinguish between informational and transformational products. While investigating the antecedent conditions of variety seeking, Van Trijp, Hoyer and Inman (1996) conclude that (1) variety seeking depends on the motivation of the buyer (intrinsic vs. extrinsic) and that (2) variety seeking is more a product category phenomenon, than a personality phenomenon (such as *need for variety seeking*). This, again, points at the importance of motivation and its close relationship to the shopped product class for understanding emotions during the purchase process.

(c) Personality. Stone (1954) finds four different types of consumer personality: economic, personalizing, ethical, and apathetic consumers. Similarly, Westbrook and Black (1985) identify six clusters, where cluster 3 (20%) and cluster 4 (10%) can be compared with Stones apathetic shoppers with low shopping motivation. Bellenger and Korgaonkar (1980) or Gentry, Grünhagen and Skupa (1997) contrast economic from recreational shoppers. It is intuitively plausible, that this overall relationship of a person to shopping activities has an influence on purchase emotions over and above the influence of product and task-related motivation.

(d) POS-factors. Frequently mentioned sources of purchase emotions are assortment, prices, atmosphere, product presentation, esthetical experiences and the quality of advice of sales clerks (Holbrook and Hirschman 1982; Jones 1999; Kraigher-Krainer 2000; Liebmann and Zentes 2001). Furthermore, specific positive or negative emotions can elicit from a broad variety of POS-related factors, reaching from the traffic and parking situation (Adaval 2001), crowding and queues (East et al. 1994), (ir)relevant product information (Esch 2001; Meyvis and Janiszewski 2002) to very specific aspects like the increasing crime rate in American shopping centers (Chaudhuri 1997).

(e) Extraneous mood effects. A relatively large amount of attention has been devoted to the influence of extraneous mood on decision processes (Bosmans and Baumgartner 2005). Starting with Schwarz and Clore's (1983) observation that respondents were more satisfied with their lives on sunny than on rainy days – but only when their attention was not drawn to the weather, the mood-as-information-hypothesis became one of the most prominent assumptions in mood research. Therefore, it can also be assumed that bad or good mood, although not valid for judging a product during a shopping trip, can influence the product related purchase evaluations, especially, if motivation for elaboration and level of cognitive engagement are low (Albarracín and Kumkale 2003) and if the product information is mood consistent (Adaval 2001).

II. Purchase emotions. As depicted in Fig. 1, these antecedents have a direct effect on purchase emotions, affecting their direction (positive or negative) and intensity (high or low). Furthermore, there is an indirect effect on purchase emotions, caused by the mediating influence of limited resources. These limited resources can either stem from limited opportunities to process (e.g. relevant information can not be accessed), or from limited abilities (e.g., time, money, cognitive capacities). Strictly speaking, such limitations need not be conceptualized separately, as intrinsic motivation (as ongoing positive emotion-chains) is interrupted by such limitations, which produce initial emotions of disappointment and subsequently result in extrinsic motivation. Correspondingly, Bagozzi, Gopinath and Nyer (1999, p. 186) state, that "... positive emotions (e.g., happiness, elation, joy) are associated with the attainment of a (sub)goal, which usually leads to a decision to continue with the plan, whereas negative emotions (e.g., frustration, disappointment, anxiety) result from problems with ongoing plans and failures to achieve desired goals ..." (similar Bosmans and Baumgartner 2005). But as (1) such limitations are of crucial importance for understanding the nature and dynamic of emotion and motivation, and (2) cybernetic processes are difficult to illustrate in static figures, I have decided to mention them separately.

Support for the assumption that intrinsic motivation calls for both the motivation *and* the capacity condition, comes from many different fields of research: Motivation research itself (Tice, Bratslavsky and Baumeister 2001; Ryan and Deci 2000); appraisal theories (Lazarus 1991; Nyer 1997; Scherer 1984); attribution theory (Heider 1958; Meyer and Försterling 1993); involvement research (Andrews, Durvasala and Akhter 1990; Poiesz and DeBont 1995); and within the field of involvement research especially from most prominent exponents of dual-process-models like Chaiken (1980; 1987), or Petty and Cacioppo (1986). Conversely, extrinsic motivation stems from *either* lacking motivation *or* lacking capacity.

Such limited resources that elicit negative purchase emotions frequently result from information overload, consumer confusion or mental inconvenience (Esch 2001; Esch and Rutenberg 2004; Fiske and Taylor 1984; Garbarino and Edell 1997; Haedrich, Hillen and Polasek 1997; Herrmann, Huber and Seilheimer 2003; Meyers-Levy and Malaviya 1999; Rudolph and Schweizer 2003; Smolowitz 2006). Whereas these negative consequences of overserving markets stem from little motivation of consumers to elaborate purchase decisions, restrictions like lacking money (Jones 1999), or time (Okada and Hoch 2004), difficulties in role interpretation (Fischer and Arnold 1990) or in understanding cultural or technical differences within complex product categories (Solomon 1986) address the capacity or the ability condition.

III. Consequences of purchase emotions. The right side of Fig. 1 depicts possible consequences of purchase emotions, derived from the literature. Although the assumption that emotions do influence decisions and behavior is well supported (Gendolla 2000), a systematic conceptualization is still missing, especially of the relationship between emotion and behavior. The first aspect, (a) perceived risk, has already been reported above.

(b) Allocation of resources. Emotions are important for allocating resources to tasks (Gross 1998). If extrinsically motivated, consumers tend to minimize efforts like preplanning, continuation of shopping trips, information search, or traveling distances (Babin, Darden and Griffin 1994; Bellenger and Korgaonkar 1980; Holbrook and Gardner 1998; Mittal and Lee 1989; Pham 1998; Westbrook and Black 1985) whereas the opposite can be expected for intrinsically motivated decisions. Following Gross (1998), Hopkinson and Pujari (1999) or, Sansone and Harackiewicz (1996), intrinsic motivation can even lead to a state, which has been termed *flow* by Csikszentmihalyi (e.g., 1975).

The most important underlying aspect seems to be the mobilization of time resources. Babin, Darden and Griffin (1994, p. 648), in their focus groups, collect statements of which one seems to best describe shopping trips under extrinsic motivation, namely to "... get in and out with a minimum of time wasted." Westbrook and Black's (1985, p. 99; 101f) apathetic shoppers are reported to have a strong motivation in "... finding exactly what I want, in the least amount of time." Correspondingly, Holbrook and Gardner (1998, p. 243) emphasize the time aspect in their definition of intrinsic vs. extrinsic motivation, stating that "...extrinsically motivated consumption should tend to cease as soon as the task is finished or the objective is attained ..." whereas "...intrinsically motivated consumer behavior should tend to continue for a longer period the more it brings pleasure, thereby maximizing the duration of enjoyable consumption experiences." p. 243 Four items in the Slama and Tashchian (1985) Purchase Involvement Scale measure time allocation.

(c) Problem elaboration. The general opinion has been for a long period, that bad mood leads to systematic processing, whereas good mood leads to heuristic processing. But this view has changed (Martin et al. 1993). Today most authors are of the opinion, that extrinsic motivation and bad mood lead to a conservative way of thinking and problem solving, adhering to existing information guided by avoidance learning and applying a bottom-up-strategy, whereas under intrinsic motivation and good mood problem solving becomes creative, guided by curiosity and a top-down-strategy and using free capacities for new solutions and creative associations (Bless et al. 1996; Cox, Cox and Zimet 2006; Fiedler and Bless 2000; Cohen and Andrade 2004). If, on the other hand, capacities (Forgas 1995) or coping potential (Nyer 1997) are restricted, people again tend to apply cognitive shortcuts like relying on feeling as information or on simplified partial search.

(d) Complaining behavior and WOM. While some authors are of the opinion, that (dis)satisfaction or regret are sufficient conditions for complaining and word-of-mouth/WOM (Reichheld 1997, 2004; Zeelenberg and Pieters 2004), others suppose that a second condition like goal relevance (Nyer 1997) or high involvement (Matzler 1997; Wirtz and Chew 2002) is necessary to produce these consequences. Anyway, Nyer and Gopinath (2005) demonstrate that vent, caused by dissatisfaction, will lead to more negative WOM, if no complaining takes place, hence leading to a stronger commitment to the articulated dissatisfaction. Derbaix and Vanhamme (2003) can show in their study, that positive vs. no vs. negative surprise lead to an increased probability of positive vs. no vs. negative WOM-behavior. But a broader discussion about other emotions than (dis)satisfaction that can cause WOM is still lacking.

(e) Variety seeking and loyalty. Unfortunately, the distinction between intrinsically and extrinsically motivated variety seeking and loyalty is still in its infancy. As a consequence, some authors suppose that bad mood always leads to less variety seeking (Kahn and Isen 1993) or that regret always leads to a higher likelihood of switching (Zeelenberg and Pieters 2004). As Fig. 1 shows, consumers are proposed to show habitual behavior if there is no schema discrepancy. In these cases the amount of variety seeking is expected to be minimized and the loyalty is expected to be maximized. This behavior can easily be misinterpreted as the effect of high satisfaction with the product, whereas it actually is the effect of lacking importance of the decision. Correspondingly, Ratner, Kahn and Kahneman (1999) use the term real variety seeking only for hedonic products and Van Trijp, Hoyer and Inman (1996) speak of *real variety seeking* if the motivation is intrinsic and of *derived varied behavior* if the motivation is extrinsic (e.g., the habitually purchased product is out of stock, which produces a schema discrepancy and a subsequent negative surprise). Similarly, in the discussion about loyalty the body of literature, distinguishing true from spurious loyalty is increasing (Bloemer and Kasper 1995; Coulter, Price and Feick 2003; Quester and Lim 2003; White und Yanamandram 2004). As Uncles, Dowling and Hammond (2003, p. 297) clearly point out, habit based loyalty can be expected for purchases, which are "…not worth the time and trouble to search for an alternative …" and that dissatisfaction will only lead to less brand loyalty, if the customer is engaged.

(f) Mood congruent retrieval, encoding, and learning effects. The idea that our perception, the retrieval of contents from memory, and the processing and storage of information tend to correspond with our momentary mood (positive vs. negative) goes back to Isen et al. (1978) and Bower (1981). There is some evidence that mood congruent information is more persuasive, if the motivation to elaborate a decision is high (Adaval 2001; DeSteno et al. 2004; Forgas 1995) and if the source of bad mood is not aware (Gorn, Goldberg and Basu 1993). But the results are far from being consistent and Bagozzi, Gopinath and Nyer (1999) summarize, that some studies show mood congruent retrieval, encoding, and learning effects, some do not.

(g) Management effects. Extrinsic shopping activities are repeatedly inevitable in consumers' everyday lives. But since they are accompanied by negative emotions they are also accompanied by the urge to cease these activities as soon as possible and not to exceed the minimum time required to do the job properly, whereas intrinsic shopping activities should produce the opposite effect and lead to a higher readiness to extend these activities. It is proposed that such *emotion management effects* influence consumers´ schemas and, subsequently, trigger the realization of limited resources (see above) in a way that, for instance, under extrinsic motivation a queue at the cashier leads to a disturbing, disappointing perturbation sooner than under intrinsic motivation. Such effects should turn up as *repair effects* for extrinsic or instrumental purchase activities and as *maintenance effects* for intrinsic or hedonic purchase activities.

Furthermore, shopping seems to play a pivotal role in the long term aspect of *mood management*. Impulsive shopping has been proposed to serve as a means for lighting up mood by distracting (Babin, Darden and Griffin 1994) and entertaining consumers, by imagining the possession of a certain product (Holbrook and Hirschman 1982), by acquiring (Coley and Burgess 2003; Hausman 2000) or using this product (e.g., simply driving somewhere with the brand new car, Bloch 1981). Tice, Bratslavsky and Baumeister (2001) are of the opinion, that in crises impulsive behavior is dominant, otherwise impulse control prevails. In crises, the urge to feel better exaggerates impulse control what they term *priority shift*. People frequently are aware of the fact that impulsive behavior is no solution in a long run and push themselves to get back to impulse control, as soon as the crisis is mastered, whereas others "… may end up paying a very high price for a better mood." (p. 65; similarly Rook 1987, Rook and Fisher 1995).

4 HYPOTHESES AND POSSIBLE DIRECTIONS OF FUTURE RESEARCH

In his legendary contribution, Levitt (1960) points at the fact that companies and customers often see the same product with different eyes. Whereas companies love their product and the technology behind it, customers do not automatically. They buy numerous products with minimum engagement and for their avoidance functionality instead of their approach functionality. Half a century later, it seems as if this eye disease, called *marketing myopia*, still has not been cured. Marketers invest in loyalty programs, luxury products, technical innovations, exaggerating assortments and the like regardless of the question whether they are promoting a hedonic or a utilitarian product, hence not fully understanding the conditions, under which these investments are valued by consumers and under which they are not valued or even punished. The following assumptions for future research and empirical foundation can be derived from the conceptualization lying before.

A set of hypotheses. Traditional Decision Theory (e.g., Becker 1976; Irle 1972; Kirsch 1977; Menges 1972) grasps man as an *unconditionally* busy and curious organism, diligently collecting information. By way of contrast, newer approaches rest on McGuire's (1969) more realistic - yet equally unsatisfactory - *lazy-organism-assumption*, claiming that man is *always* lazy and miserly with his resources (Chaiken 1980; Fiske and Taylor 1984; Forgas 1995, 2000; Petty and Cacioppo 1986). It seems plausible to guess that the former approach talks of the intrinsically motivated man, whereas the latter talks of the extrinsically motivated man. Anyhow, the varying readiness of consumers to allocate resources like time, money, or cognitive capacity to decision tasks depending on motivation has numerous consequences that can be hypothesized in the following way:

1. Extrinsically motivated shoppers are less interested in luxury products, technical sophistication, or exaggerating assortments than intrinsically motivated shoppers.

2. Shopping for intrinsic products leads to a higher readiness to adopt product innovations of this category relatively early whereas extrinsically motivated shopping

leads to a relatively late adoption of innovations.

3. Intrinsically motivated buyers are heavy users of all kinds of information, especially if the perceived risk is high, compared to extrinsically motivated buyers who prefer the more convenient sources like recommendations and avoid sources that require cognitive effort like reports in newspapers.

As has been conceptualized, an intrinsic shopping trip can and will convert to an extrinsic one if the shopper experiences irresolvable limitations in opportunity and/or ability to process. Limitations in *opportunity* entail experiences like missing or unsuitable information, lacking competence of sales clerks, or dissatisfying POS-conditions ranging from assortment, prices, product presentation to the traffic situation. Limitations in *ability* entail restrictions regarding money (e.g., limited budget), time (e.g., time pressure), and cognitive capacity (e.g., lacking capacity to understand the technical features of a TV-set).

4. If an intrinsically motivated shopper experiences capacity limitations or lacking opportunities which cannot be resolved, the shopping motivation will convert from intrinsic to extrinsic and the customer will end up either giving up the purchase goal or purchasing in an extrinsically motivated way.

The time problem is of special importance for understanding consumer behavior. As mentioned above intrinsic activities are inherently pleasurable. The hedonic-man-assumption leads to the conclusion that intrinsically motivated shoppers are in a state of positive mood and will apply a *maintenance-strategy*, whereas the extrinsically motivated are expected to apply a *repair strategy*. It has been proposed to understand *time as a resource as is money* (Okada and Hoch 2004, p. 322). From an intrinsic-extrinsic point of view it looks different: Time under extrinsic motivation is a means to an end, hence investment. Anyway, time under intrinsic motivation is – at least partly and as long as there are no time limitations (see above) - an end in itself. The exact role of time in shopping trips is far from being resolved and interacts with motivation and capacity in a complex manner (Suri and Monroe 2003). But the following assumption seems plausible:

5. Extrinsically motivated customers are more likely to appreciate offers that help them minimize the time wasted during a shopping trip (short journey there, small and competent assortment of key products, high probability of finding exactly what they are searching for, convenient risk reduction opportunities provided by key information, confident sales clerks, different kinds of guarantees etc.) while intrinsically motivated customers appreciate opportunities to extend the shopping trip (large assortments, esthetic impressions, competent sales clerks and experts to exchange experiences about pros and cons, locations for hanging around, support for the social and the fantasy aspect of shopping etc.).

Intrinsic vs. extrinsic motivation have been conceptualized only to appear if there is an initial emotion, triggered by schema discrepancy and subsequent surprise. Habitual behavior is a very – if not the most important and dominant – behavioral tendency of consumers (Grünhagen, Grove and Gentry 2003) and entails habits like shopping days, shopping hours, routes through town, family routines, banking routines, employment habits and the like. Consequently, consumers are (at least spuriously) loyal not only to brands, but also to product classes, price categories, retailer types, shops, sales clerks and other entities. Understanding variety seeking and loyalty is closely intertwined with understanding this network of daily routines and habits. Hence, a distinction has to be made between real variety seeking, resp. loyalty and derived varied behavior, resp. switching behavior.

6. If there is no initial surprise (positive vs. negative) shoppers will buy the product they bought last time they were shopping for that item. This behavior can be called *spurious loyalty*, lacking any variety seeking. A nega-

tive surprise (e.g., product out of stock, customer recalls that the item purchased last time performed poorly, information that the product bears specific risks etc.) leads to extrinsic motivation. Subsequently, *derived varied behavior* and brand switch can be expected. Finally, if a positive surprise appears (e.g., the decision is experienced as pleasurable, an advantageous cue enters the perception system – a price cut, a new taste or smell, a promotion etc.) the shopper is intrinsically motivated. He now is motivated to restrictively or fully check the assortment (*real variety seeking*) and may end up returning to the brand bought earlier (*real brand loyalty*) or switch.

Whereas there is much research on (dis)satisfaction-triggered word-of-mouth (WOM), comparably little is known about WOM on hedonic vs. utilitarian products (Childers and Rao 1992). It seems obvious that emotion-elicited WOM is far more than just telling others about his or her (dis)satisfaction. One will hardly find people discussing the pros and cons of detergents or batteries, but one will find them talking about cars, holiday journeys, or mobile phones, because these products are of intrinsic nature for most customers (Kraigher-Krainer 2000). Furthermore, WOM is considered to be an important risk reduction strategy (Bauer 1960; Gemünden 1985; Mitchell 1999).

7. There is more WOM about intrinsic products and services than about extrinsic products and services.

8. Intrinsically motivated purchases lead to a higher probability that the shopper will perceive himself as opinion leader, whereas extrinsically motivated purchases lead to a higher probability that the shopper will perceive himself as opinion seeker.

9. Extrinsically motivated shopping leads to a higher tendency to delegate the decision to experts and credible others and to rely on their recommendations, especially if the perceived risk is high, whereas intrinsically motivated shopping leads to a higher tendency to come to a decision on one's own expertise.

If in a bad mood, shoppers might use pleasurable shopping trips as a repair-strategy. It can be assumed that extrinsic shopping trips and products do not have the emotional capacity to serve this goal. As Stern (1962) has clearly pointed out, a pure impulse product has to meet two conditions, namely (1) it is accompanied by positive emotions and (2) the stake is low. Hence, whereas intrinsic products in general should bear the potential of distracting people from their bad mood or helping them on mood repair, regardless, whether the shopping trip culminates in a purchase or just serves as stimulating fantasies and daydreams, intrinsic products with low perceived risk should be especially suitable for mood repair.

10. Intrinsic products and services bear a higher potential in helping shoppers to overcome bad mood than extrinsic products and services.

11. The higher the intrinsic value and the lower the perceived risk, the higher the probability of purchasing a product or service without preplanning.

5 MANAGERIAL IMPLICATIONS

Managing extrinsic products. Whereas investments in big assortment, luxury products, and new technology might be good investments for intrinsic products, managers have to be careful with extrinsic products as the probability is high that these investments are neither perceived nor worshipped by the consumer. Even worse, such investments can cause damage for the company as they can unnecessarily complicate extrinsic shopping trips – the customer might decide for the easier solution provided by the competitor. He lacks curiosity about the company's beloved offers. Hofer/Aldi is often cited for their low prices. But they should also be cited for their restricted assortment (at least in the past). I believe that consumers value the way Hofer/Aldi keep their assortment scanty: They have tested and routed out all the unnecessary stuff in advance, hence provide convenience by an intelligent assortment. They work like consumer reports. Best price is also some kind of convenience. No annoying comparison necessities or cognitive dissonance after purchase if the company has the image of reliably having an unbeaten price-quality-ratio. They offer outlets in the vicinity of the customer, a slim information policy, and sufficient parking space. Taken together they help save time, money and mental capacity. I think the best describing strategy for companies that sell extrinsic products of low risk is Treacy and Wiersema's (1995) *operational excellence*.

The situation for extrinsic high risk products is a different one. Remember that extrinsic purchases follow the principle of *in-and-out-with-a-minimum-of-time-wasted*. For low risk products this repair strategy is applicable, but not for high risk products, as the high stake could cause subsequent negative emotions like regret (Herrmann, Huber and Seilheimer 2003), dissonance-related emotions (Harmon-Jones 2000), or guilt (Ruth, Brunel and Otnes 2002). As hypothesized earlier, a good way to overcome an unpleasant, yet complicating decision, is to let the decision rest on the opinion of competent and trustworthy others. Again it is Treacy and Wiersema (1995) who offer a strategy – *customer intimacy strategy* – that properly describes the management of extrinsic high risk products or services. Such competent and trustworthy others are ideally friends, relatives, and other kinds of unpaid opinion leaders (e.g., Flynn, Goldsmith and Eastman 1996); market mavens (e.g., Feick and Price 1987); agents (e.g., West 1996); or surrogates (e.g., Solomon 1986), as they have the highest credibility. But they often lack competence which brings into play the company-run sales person or consultant. In the case of extrinsic high risk products consumers show a higher preparedness to pay for competent advice and seem to prefer the expertise of sales clerks in the more expensive specialty shop compared to the poor-service-box-mover (whereby I admit: clever customers sequentially use both, which also has to be managed in the long run).

Unfortunately, little is known about the question of how WOM can be actively managed by the company (Wirtz and Chew 2002), although many, especially the small of them, often depend solely on WOM (Stokes and Lomax 2002) for their image-building. Additionally, through the internet the diffusion of WOM has accelerated and will further accelerate (Henning-Thurau 2004) which points at the necessity to dramatically increase research activities in this field of marketing.

Managing intrinsic products. To start with the WOM aspect it has been hypothesized, that intrinsically motivated buyers are heavy users of all kinds of information if they perceive high risk. They engage in WOM, seek other experts, carefully read trade journals, consumer reports and articles in newspapers, listen to all kinds of advertising, and curiously scan the POS for new solutions. They are experts themselves and this is why serving these market segments requires excellent products and employees – Treacy and Wiersema (1995) calls the adequate strategy *product leadership*. Over and above their importance as customers, these segments may serve as opinion leaders for the above mentioned extrinsically motivated shoppers. This makes them key players in the information diffusion process and this is a second reason to aim for excellence in product quality, technology, and service, regardless of whether they are targeted as customers. Furthermore, as they are the most probable early adopters, they also play a pivotal role in the introduction stage of innovations.

Intrinsically motivated shoppers enjoy their trip. So there is no hurry, they do not have to repair mood; their aim is rather to maintain mood. All kinds of adventurous shopping and all methods of prolonging the trip are welcome as long as there are no limitations in capacity. It is important to understand that such limitations can seriously undermine intrinsic purchases. Managing intrinsic products therefore also entails the management of these limitations. Some of them can hardly be managed – e.g., time limitations. Others can, by offering professional information systems, a variety of brand new products, real experiences, customer friendly loans – just to name a few of them.

Finally, it seems well supported that intrinsic shopping – especially the shopping for low risk items – is an appropriate and frequently applied means for mood repair. It's up to the marketer whether he is interested in actively addressing people in a bad mood or not - due to ethical reservations.

Managing surprise: Emotions have been conceptualized to appear if and only if some expectation is disconfirmed and positive or negative surprise appears. This is good news for companies with high market shares. They have to improve the *management of no surprise*. And they have to deal with the opportunities of spurious loyalty and the threads of derived varied behavior. Anyway, for the rival it means learning the management of *(positive or negative) surprise* (see Ansoff 1976; Derbaix/Vanhamme 2003) which might bear the opportunity for the customer's reorientation, eventually leading to real variety seeking and brand switching behavior. But research in this field is still in its infancy and it would be too much to further explicate that question here.

The aim of this paper has been, to track down the antecedents and consequences of purchase emotions. Deci and Ryan's *Self-Determination-Theory* has turned out to be a powerful tool to explain and better understand such purchase emotions. It can be stated that a better understanding of the hedonic and utilitarian value of products and services can be a good starting point with high explanatory power for both marketing scholars and marketing practitioners.

6 REFERENCE

[1] Aaker,J.L. and Lee,A.Y. (2001) "I" Seek Pleasure and "We" Avoid Pains: The Role of Self-Regulatory Goals in Information Processing and Persuasion. *Journal of Consumer Research*, **28/6**, 33–49.

[2] Adaval,R. (2001) Sometimes It Just Feels Right: The Differential Weighting of Affect-Consistent and Affect-Inconsistent Product Information. *Journal of Consumer Research*, **28/June**, 1–17.

[3] Albarracín,D. and Kumkale,G.T. (2003) Affect as Information in Persuasion: A Model of Affect Identification and Discounting. *Journal of Personality and Social Psychology*, **84/3**, 453–469.

[4] Andrade,E.B. (2005) Behavioral Consequences of Affect: Combining Evaluative and Regulatory Mechanisms. *Journal of Consumer Research*, **32/Dec.**, 355–362.

[5] Andrews,J.C., Durvasala,S. and Akhter,S.H. (1990) A Framework for Conceptualizing and Measuring the Involvement Construct in Advertising Research. *Journal of Advertising*, **19/4**, 27–40.

[6] Ansoff,H.I. (1976) Managing Surprise and Discontinuity – Strategic Response to Weak Signals. *Schmalenbachs Zeitschrift für betriebswirtschaftliche Forschung*, **28/3**, 129–130.

[7] Babin,B.J., Darden,W.R. and Griffin,M. (1994) Work and/or Fun: Measuring Hedonic and Utilitarian Shopping Value. *Journal of Consumer Research*, **20/4**, 644–656.

[8] Bagozzi,R.P., Gopinath,M. and Nyer,P.U. (1999) The Role of Emotions in Marketing. *Journal of the Academy of Marketing Science*, **27/2**, 184–206.

[9] Batra,R. and Ahtola,O.T. (1990) Measuring the Hedonic and Utilitarian Sources of Consumer Attitudes. *Marketing letters*, **2/2**, 159–170.

[10] Bauer,R.A. (1960) *Consumer behavior as Risk Taking. Risk Taking and Information Handling in Consumer Behavior.* In: Cox,D.F., Cambridge Mass, Harvard University Press, 389–398.

[11] Becker,G.S. (1976) *Der ökonomische Ansatz zur Erklärung menschlichen Verhaltens.* Mohr, Tübingen.

[12] Bellenger,D.N. and Korgaonkar,P.K. (1980) Profiling the Recreational Shopper. *Journal of Retailing*, **56/3**, 77–92.

[13] Bless,H., Bohner,G. and Schwarz,N. (1991) Gut gelaunt und leicht beeinflussbar? Stim-

mungseinflüsse auf die Verarbeitung persuasiver Kommunikation. *Psychologische Rundschau*, **43**, 1–17.

[14] Bless,H., Clore,G., Schwarz,N., Golisano,V., Rabe,C. and Wölk,M. (1996) Mood and the use of scripts: Does happy mood make people really mindless? *Journal of Personality and Social Psychology*, **71**, 665–679.

[15] Bloch,P.H. (1981) An Exploration into the scaling of Consumers´ Involvement with a product class. *Advances in Consumer Research*, **6**, 61–65.

[16] Bloemer,J.M.M. and Kasper,H.D.P. (1995) The complex relationship between consumer satisfaction and brand loyalty. *Journal of Economic Psychology,* **16**, 311–329.

[17] Bosmans,A. and Baumgartner,H. (2005) Goal-Relevant Emotional Information. When Extraneous Affect Leads to Persuasion and When It Does Not. *Journal of Consumer Research*, **32/Dec.**, 424–34.

[18] Bower,G. (1981) Mood and Memory. *American Psychologist*, **36**, 129–148.

[19] Brown,S.P., Homer,P.M. and Inman,J.J. (1998) A Meta-Analysis of Relationships Between Ad-Evoked Feelings and Advertising Responses. *Journal of Marketing Research*, **35/Feb.**, 114 –126.

[20] Chaiken,S. (1980) Heuristic Versus Systematic Information Processing and the Use of Source Versus Message Cues in Persuasion. *Journal of Personality and Social Psychology*, **39**, 752–766.

[21] Chaiken,S. (1987) *The Heuristic Model of Persuasion. In: Zanna,M.P., Olson,J.M. and Herman,C.P. (Eds.) Social Influence*: *The Ontario Symposium.* 5, Erlbaum, New York, 3–39.

[22] Chaudhuri,A. (1997) Consumption Emotion and Perceived Risk: A Macro-Analytic Approach. *Journal of Business Research*, **39/2**, 81–92.

[23] Childers,T.L. and Rao,A.R. (1992) The Influence of Familial and Peer - based Reference Groups on Consumer Decisions. *Journal of Consumer Research*, **19/2**, 198–211.

[24] Ciompi,L. (1998) *Affektlogik. Über die Struktur der Psyche und ihre Entwicklung.* Klett-Cotta, Stuttgart.

[25] Cohen,J.B. and Areni,C.S. (1991) *Affect and Consumer Behavior. In: Handbook of Consumer Behavior. Robertson,T.S. and Kassarjian,H.H. (Eds.)* Prentice Hall, Englewood Cliffs, 188–240.

[26] Cohen,J.B. and Andrade,E.B. (2004) Affective Intuition and Task-Contingent Affect Regulation. *Journal of Consumer Research*, **31/Sept.**, 358–367.

[27] Coley,A. and Burgess,B. (2003) Gender differences in cognitive and affective impulse buying. *Journal of Fashion Marketing and Management*, **7/3**, 282–295.

[28] Conchar,M.P., Zinkhan,G.M., Peters,C., and Olavarrieta,S. (2004) An Integrated Framework for the Conceptualization of Consumers' Perceived-Risk Processing. *Journal of the Academy of Marketing Science*, **32**, 418–436.

[29] Coulter,R.A., Price,L.L., and Feick,L. (2003) Rethinking the Origins of Involvement and Brand Commitment: Insights from Postsocialist Central Europe. *Journal of Consumer Research*, **30/Sept.**, 151–169.

[30] Cox,A.D., Cox,D. and Zimet,G. (2006) Understanding Consumer Responses to Product Risk Information. *Journal of Marketing,* **70/Jan.**, 79–91.

[31] Cszikszentmihalyi,M. (1975) *Beyond boredom and anxiety.* Jossey-Bass, San Francisco.

[32] Damasio,A.R. and Damasio,H. (1994) *Sprache und Gehirn. In: Singer W.(Ed.) Gehirn und Bewusstsein.* Spektrum, Heidelberg, 58–66.

[33] Deci,E.L. and Ryan,R.M. (1985) *Intrinsic Motivation and Self-Determinant. In: Human Behavior.* Plenum, New York.

[34] Derbaix,C., Vanhamme,J. (2003) Inducing word-of-mouth by eliciting surprise - a pilot investigation. *Journal of Economic Psychology*, **24**, 99–116.

[35] Desteno,D., Petty,R.E., Rucker,D.D., Wegener,D.T. and Braverman,J. (2004) Discrete Emotions and Persuasion: The Role of Emption-Induced Expectancies. *Journal of Personality and Social Psychology*, **86/1**, 43–56.

[36] Dowling,G.R. and Staelin,R. (1994) A Model of Perceived Risk and Intended Risk-handling Activity. *Journal of Consumer Research*, **21/June**, 119–134.

[37] East,R., Lomax,W., Willson,G. and Harris,P. (1994) Decision Making and Habit in Shopping Times. *European Journal of Marketing*, **28/4**, 56–71.

[38] Esch,F.R. (2001) *Wirkung integrierter Kommunikation. Ein verhaltenswissenschaftlicher Ansatz für die Werbung.* **3,** Deutscher Univ.Verlag, Wiesbaden.

[39] Esch,F.R., Rutenberg,J. (2004) Mental Convenience beim Einkaufen. *Thexis* **4/4,** 22–26.

[40] Feick,L., Price,L.L. (1987) The Market Maven: A Diffuser of Marketplace Information. *Journal of Marketing,* **51/Jan.,** 83–97.

[41] Fiedler,K. and Bless,H. (2000) *The formation of beliefs at the interface of affective and cognitive processes. In: Frijda,N.H., Manstead,S.R. and Bem.S. (Eds.) Emotions and Beliefs. How Feelings Influence Thoughts.* Cambridge University Press, Cambridge, 144–170.

[42] Fischer,E., Arnold,S.J. (1990) More than a Labor of Love: Gender Roles and Christmas Gift Shopping. *Journal of Consumer Research,* **17/Dec.,** 333–345.

[43] Fiske,S.T. and Taylor,S.E. (1984) *Social Cognition.* Addison-Wesley, Reading.

[44] Flynn,L.R., Goldsmith,R.E., Eastman,J.K. (1996) Opinion Leaders and Opinion Seekers: Two New Measurement Scales. *Journal of the Academy of Marketing Science,* **24/2,** 137–147.

[45] Forgas,J.P. (1995) Mood and Judgment: The Affect Infusion Model (AIM). *Psychological Bulletin,* **117/1,** 39–66.

[46] Forgas,J.P. (2000) *Feeling is Believing? The role of processing strategies in mediating affective influences on beliefs. In: Frijda,N.H., Manstead,S.R. and Bem.S. (Eds.) Emotions and Beliefs. How Feelings Influence Thoughts.* Cambridge University Press, Cambridge, 108–143.

[47] Frijda,N.H., Manstead,S.R. and Bem,S. (2000) *The influence of emotions on beliefs. In: Frijda,N.H., Manstead,S.R. and Bem.S. (Eds.) Emotions and Beliefs. How Feelings Influence Thoughts.* Cambridge University Press, Cambridge, 1–9.

[48] Garbarino,E.C., Edell,J.A. (1997) Cognitive Effort, Affect, and Choice. *Journal of Consumer Research,* **24/Sept.** 147–158.

[49] Gemünden,H.G. (1985) Perceived risk and information search. A Systematic meta-analysis of the empirical evidence. *International Journal of Research in Marketing,* **2,** 79–100.

[50] Gendolla,G.H.E. (2000) On the impact of Mood on Behavior: An Integrative Theory and a Review. *Review of General Psychology,* **4/4,** 378–408.

[51] Gentry,J.W., Grünhagen,M. and Skupa,L.J. (1997) *Insight into the likely consumer response to expanded retail hours globally: an investigation of the emic meaning of Saturday versus Sunday shopping. In: King,R.L. (Ed.) Proceedings of the Fifth Triennial AMS/ACRA National Retailing Conference. Special Conference Series Vol. VII., "Retailing: End of a century and a look to the future",* 8, AMS, St. Louis, 26–30.

[52] Glasersfeld,E.V. (1997) *Radikaler Konstruktivismus. Ideen, Ergebnisse, Probleme.* Suhrkamp, Frankfurt/Main.

[53] Gorn,G.J., Goldberg,M.E. and Basu,K. (1993) Mood, Awareness, and Product Evaluation. *Journal of Consumer Psychology,* **2/3,** 237–256.

[54] Gorn,G., Pham,M.T. and Sin,L.Y. (2001) When Arousal Influences Ad Evaliation and Valence Does Not (and Vice Versa). *Journal of Consumer Psychology,* **11/1,** 43–55.

[55] Gross,J.J. (1998) The Emerging Field of Emotion Regulation: An Integrative Review. *Review of General Psychology,* **2/3,** 271–299.

[56] Grünhagen,M., Grove,S.J., Gentry,J.W. (2003) The dynamics of store hour changes and consumption behavior. Results of a longitudinal study of consumer attitudes toward Saturday shopping in Germany. *European Journal of Marketing,* **37/11-12,** 1810–1817.

[57] Haedrich,H., Hillen,H. and Polasek,S. (1997) *Neue Ansprüche des Kunden an eine bedarfsgerechte Kommunikation. Belz Christian (Ed.), Strategisches Direct Marketing,* Ueberreuter, Korneuburg, 73–100.

[58] Hänze,M. (1998) *Denken und Gefühl. Wechselwirkung von Emotion und Kognition im Unterricht.* Luchterhand, Neuwied.

[59] Harmon-Jones,E. (2000) *A cognitive dissonance theory perspective on the role of emotion in the maintenance and change of beliefs and attitudes. In: Frijda,N.H., Manstead,A.S.R. and Bem,S. (Eds.) Emotions and Beliefs. How Feelings Influence Thoughts.* Cambridge University Press, Cambridge, 185–211.

[60] Hausman,A. (2000) A multi-method investigation of consumer motivations in impulse buying behaviour. *Journal of Consumer Marketing,* **17/5,** 403–419.

[61] Heider,F. (1958) *The Psychology of Interpersonal Relations.* Wiley Sons, New York.

[62] Henning-Thurau,T. (2004) "Word-of-Mouse" - Warum Kunden anderen Kunden im Internet zuhören, *Jahrbuch der Absatz- und Verbraucherforschung,* **50,** 52–75.

[63] Herrmann,A., Brandenberg,A., Lyczek,B. und Schaffner,D. (2004) Wahrnehmungswerte als Herausforderung für die Messung der Marketingproduktivität - Grenzen vorhandener Ansätze und Vorschlag eines Synthesemodells. *Thexis,* **3,** 2–7.

[64] Herrmann,A., Huber,F. und Seilheimer,C., (2003) Die Qual der Wahl: Die Bedeutung des Regret bei Kaufentscheidungen. *Schmalenbachs Zeitung für betriebswirtschaftliche Forschung,* **55/5,** 224–249.

[65] Holbrook,M.B. (1986) *Emotion in the Consumption Experience. Toward A New Model of the Human Consumer. The Role of Affect. In: Consumer Behavior, Robert A. Peterson, Wayne D. Hoyer, and William R. Wilson (Eds.)* Lexington Books, Lexington MA, 17–52.

[66] Holbrook,M.B., Gardner,M.P. (1998) How Motivation Moderates the Effects of Emotions on the Duration of Consumption. *Journal of Business Research,* **42/3,** 241–252.

[67] Holbrook,M.B., Hirschman, E.C. (1982) The Experiential Aspects of Consumption: Consumer Fantasies, Feelings, and Fun. *Journal of Consumer Research,* **9/Sept.,** 132–140.

[68] Hopkinson,G.C. and Pujari,D. (1999) A factor analytic study of the sources of meaning in hedonic consumption. *European Journal of Marketing,* **33/3-4,** 273–290.

[69] Irle,M. (1972) *Voraussetzungen und Strukturen der Entscheidung. Kurzrock,R. (Ed.) Forschung und Information, Bd. 12, Systemtheorie,* Colloquium, Berlin, 170–177.

[70] Isen,A.M. and Geva,N. (1987) The influence of Positive Affect on Acceptable Level of Risk: The Person with a Large Canoe Has a Large Worry. *Organizational Behavior and Human Decision Processes,* **39,** 145–54.

[71] Isen,A.M. and Simmonds,S.F. (1978) The Effect of Feeling Good on a Helping Task that Is Incompatible with Good Mood. *Social Psychology,* **41/4,** 346–9.

[72] Isen,A.M., Shalker,T.E., Clark,M. and Karp,L. (1978) Affect, Accessibility of Material in Memory, and Behavior: A Cognitive Loop? *Journal of Personality and Social Psychology,* **36/Jan.,** 1–12.

[73] Izard,C.E. (1991) *The Psychology of Emotions.* Plenum Press, New York.

[74] Jones,M.A. (1999) Entertaining Shopping Experiences: An Exploratory Investigation. *Journal of Retailing and Consumer Services,* **6/3,** 129–139.

[75] Kahn,B.E. und Isen,A.M. (1993) The Influence of Positive Affect on Variety Seeking among Safe, Enjoyable Products. *Journal of Consumer Research,* **20/Sept.,** 257–270.

[76] Kirsch,W. (1977) *Das Modell des Homo oeconomicus. KIRSCH Werner (Ed.) Einführung in die Theorie der Entscheidungsprozesse,* Gabler, Wiesbaden, 27–60.

[77] Kleinginna,A.M. and Kleinginna,P.R. (1981) A Categorized List of Emotion Definitions, with Suggestions for a Consensual Definition. *Motivation and Emotion,* **5/4,** 345–379.

[78] Kraigher-Krainer,J. (2000) *Käuferverhaltenstypen im Spannungsfeld realer und virtueller Marktplätze. In: Foscht Thomas, Jungwirth Georg, Schnedlitz Peter (Eds.) Zukunftsperspektiven für das Handelsmanagement, Konzepte - Instrumente –Trends.* Deutscher Fachverlag, Frankfurt am Main, 386–410.

[79] Kraigher-Krainer,J. (2005) *Das ECID-Modell. Konzeptualisierung, Operationalisierung, empirische Prüfung und strategische Iimplikationen.* Habilitationsschrift, Graz.

[80] Laros,F.J.M. and Steenkamp,J.B.E.M. (2005) Emotions in consumer behavior: a hierarchical approach. *Journal of Business Research,* **58,** 1437–1445.

[81] Lazarus,R.S. (1991) *Emotion and Adaption.* Oxford University Press, New York.

[82] LeDoux,J. (1996) *The emotional brain. The mysterious underpinnings of emotional life.* Touchstone, New York.

[83] Lerner,J.S. and Keltner,D. (2000) Beyond valence: Toward a model of emotion-specific influences on judgement and choice. *Cognition and Emotion,* **14/4,** 473–493.

[84] Lerner,J.S., Small,D.A. and Loewenstein,G. (2004) Heart Strings and Purse Strings. Carry-over Effects of Emotions on Economic Decisions. *Psychological Science,* **15/5,** 337–341.

[85] Levitt,T. (1960) *Marketing Myopia, wiederabgedruckt. In: Enis,B.M., and Cox,K.K, (Eds.) (1991): Marketing Classics. A Selection of In-*

fluentual Articles 7, Prentice Hall, Englwood Cliffs, 3–21.

[86] Liebmann,H.-P. and Zentes,J. (2001) *Handelsmanagement*, Vahlen, München.

[87] Martin,L.L., Ward,D.W., Achee,J.W. and Wyer,R.S. (1993) Mood as Input: People Have to Interpret the Motivational Implications of Their Moods. *Journal of Personality and Social Psychology,* **64/3**, 317–326.

[88] Matzler,K. (1997) *Kundenzufriedenheit und Involvement 1*, DeutscherUnivVerlag, Wiesbaden.

[89] McGuire,W.J. (1969) The Nature of Attitudes and Attitude Change. In: Lindzey,G., Aronson,E. (Eds.) *The Handbook of Social Psychology*, 2/3, Addison-Wesley, Reading, 136–314.

[90] Menges,G. (1972) *Entscheidungsmodelle in den Wirtschaftswissenschaften. In: Kurzrock,R. (Ed.) Forschung und Information, Bd. 12. Systemtheorie*. Colloquium, Berlin.

[91] Meyer,W.-U. und Försterling,F. (1993) *Die Attributionstheorie. In: Frey,D. und Irle,M. (Ed.) Theorien der Sozialpsychologie, Bd. I: Kognitive Theorien*. Huber, Bern, 175–214.

[92] Meyers-Levy,J., Malaviya,P. (1999); Consumers Processing of Persuasive Advertisements: An Integrative Framework of Persuasion Theories. *Journal of Marketing*, **63**, 45–60.

[93] Meyvis,T., Janiszewski,C. (2002) Consumers´ Beliefs about Product Benefits: The Effect of Obviously Irrelevant Product Information. *Journal of Consumer Research*, **28/3**, 618–635.

[94] Mitchell,V.-W. (1999) Consumer Perceived Risk: Conceptualisations and models. *European Journal of Marketing*, **33/1-2**, 163–195.

[95] Mittal,B. (1987) *A Framework for Relating Consumer Involvement to Lateral Brain Functioning. In: Advances in Consumer Research, 14. Wallendorf,M. and Anderson,P. (Eds.) Provo, UT: Association for Consumer Research*, 41–45.

[96] Mittal,B. and Lee,M.-S. (1989) A Causal Model of Consumer Involvement. *Journal of Economic Psychology*, **10**, 363–389.

[97] Morse,G. (2006) Decision and Desire. *Harvard Business Review,* **Jan.**, 42–51.

[98] Nyer,P.U. (1997) A study of the Relationship Between Cognitive Appraisals and Consumption Emotions. *Journal of the Academy of Marketing Science*, **25/4**, 296–304.

[99] Nyer,P.U. and Gopinath,M. (2005) Effects of Complaining Versus Negative Word of Mouth on Subsequent Changes in Satisfaction, The Role of Public Commitment. *Psychology and Marketing*, **22/12**, 937–953.

[100] Nygren,T.E., Isen,A.M., Taylor,P.J. and Dulin,J. (1996) *The Influence of Positive Affect on the Decision Rule in Risk Situations: Focus on Outcome (and Especially Avoidance of Loss) Rather Than Probability. Organizational Behavior and Human Decision Processes,* **66/1**, 59–72.

[101] Okada,E.M., Hoch,S.J. (2004) Spending Time vs. Spending Money. *Journal of Consumer Research,* **31/Sept.**, 313–323.

[102] Olds,J., Milner,P. (1954) Positive Reinforcement Produced by Electrical Stimulation of Septal Area and Other Regions of Rat Brain. *Journal of Comparative Physiological Psychology,* **47**, 419–427.

[103] Oliver,R.L. (1980) A Cognitive Model of the Antecedents and Consequences of Satisfaction Decisions. *Journal of Marketing Research*, **17/Nov.**, 460–469.

[104] Petty,R.E. and Cacioppo,J.T. (1986) *Communication and Persuasion: Central and Peripheral Routes to Attitude Change*. Springer, New York.

[105] Pham,M.T. (1998) Representativeness, Relevance, and the Use of Feelings in Decision Making. *Journal of Consumer Research*, **25/Sept.**, 144–159.

[106] Poiesz,T.B.C. and Bont,C.J.PM. (1995) *Do We Need Involvement to Understand Consumer Behavior? Advances in Consumer Research 22, In: Kardes,F. und Sujan,M. (Eds.) Provo: Association for Consumer Research*, 448–52.

[107] Powers,W.T. (1973) Feedback: Beyond Behaviorism. Stimulus-response laws are wholly predictable within a control-system model of behavioral organization. *Science*, **179**, 351–356.

[108] Quester,P., Lim,A.L. (2003) Product involvement/brand loyalty: is there a link? *Journal of Product & Brand Management*, **12/1**, 22–38.

[109] Ratner,R.K., Kahn,B.E. und Kahneman,D. (1999) Choosing Less-Preferred Experiences for the Sake of Variet. *Journal of Consumer Research,* **26/June,** 1–15.

[110] Reichheld,F.F. (1997) *Der Loyalitätseffekt, die verborgene Kraft hinter Wachstum, Gewinnen und Unternehmenswert.* Campus, Frankfurt/Main.

[111] Reichheld,F.F. (2004) Mundpropaganda als Maßstab für den Erfolg. *Harvard Business Manager,* **3,** 22–35.

[112] Rook,D.W. (1987) The Buying Impulse. *Journal of Consumer Research,* **14/Sept.,** 189–199.

[113] Rook,D.W., Fisher,R.J. (1995) Normative influences on Impulse Buying Behavior. *Journal of Consumer Research,* **22/Dec.,** 305–313.

[114] Rossiter,J.R., Percy,L. and Donovan,R.J. (1991) A Better Advertising Planning Grid. *Journal of Advertising Research,* **Oct./Nov.,** 11–21.

[115] Rudolph,T. and Schweizer,M. (2003) Kunden wieder zu Käufern machen. *Harvard Business Manager,* **2/03,** 23–33.

[116] Ruth,J.A., Brunel,F.F.B. and Otnes,C.C. (2002) Linking Thoughts to Feelings: Investigating Cognitive Appraisals and Consumption Emotions in a Mixed-Emotions Context. *Journal of the Academy of Marketing Science,* **30/1,** 44–58.

[117] Ryan,R.M. and Deci,E.L. (2000) Intrinsic and Extrinsic Motivations: Classic Definitions and New Directions. *Contemporary Educational Psychology,* **25/1,** 54–67.

[118] Sansone,C. and Harackiewicz,J.M. (1996) *"I don't feel like it": The Function of Interest in Self-Regulation.* In: Martin,L.L. and Tesser,A. (Eds.) *Striving and feeling: Interactions among goals, affect, and self regulation.* Erlbaum, Hillsdale, 203–228.

[119] Scherer,K.R. (1984) *On the nature and function of emotion: A component process approach.* In: Scherer,K.R. and Ekman,P. (Eds.) *Approaches to emotion.* Erlbaum, Hillsdale, 293–318.

[120] Scherer,K.R. (1995) Plato´s legacy: Relationships between cognition, emotion, and motivation. *Geneva Studies in Emotion and Communication,* **9/1,** 1–7.

[121] Schmalen,H. (1994) Das hybride Kaufverhalten und seine Konsequenzen für den Handel. *Zeitschrift für Betriebswirtschaft (ZfB),* **64/10,** 1221–1240.

[122] Schwarz,N. and Clore,G.L. (1983) Mood, Misattribution, and Judgments of Well-Being: Informative and Directive Functions of Affective States. *Journal of Personality and Social Psychology,* **45/3,** 513–523.

[123] Schwarz,N. and Clore,G.L. (2003) Mood as Information: 20 Years later. *Psychological Inquiry,* **14/3-4,** 296–303.

[124] Sherry,J.F., McGrath,M.A. and Leyv,S.J. (1993) The Dark Side of the Gift. *Journal of Business Research,* **28/Nov.,** 225–244.

[125] Slama,M.E. and Tashchian,A. (1985) Selected Socioeconomic and Demographic Characteristics Associated with Purchasing Involvement. *Journal of Marketing,* **49,** 72–82.

[126] Smolowitz,I. (2006) Consumer Marketing: A Flawed Strategy. *Marketing Matters Newsletter,* **5/6,** AMA, Chicago.

[127] Solomon,M.R. (1986) The Missing Link: Surrogate Consumers in the Marketing Chain. *Journal of Marketing,* **50/Oct.,** 208-218.

[128] Stern,H. (1962) The Significance of Impulsive Buying Today. *Journal of Marketing,* **26/2,** 59–62.

[129] Stokes,D. and Lomax,W. (2002) Taking control of word of mouth marketing: the case of an entrepreneurial hotelier. *Journal of Business and Enterprise Development,* **9/4,** 349–357.

[130] Stone,G.P. (1954) City Shoppers and Urban Identification: Observations on the Social Psychology of City Life. *American Journal of Sociology,* **60,** 36–45.

[131] Suri,R. and Monroe,K.B. (2003) The effects of time constraints in Consumers´ Judgements of Prices and Products. *Journal of Consumer Research,* **30/June,** 92–104.

[132] Tice,D.M., Bratslavsky,E. and Baumeister,R.F. (2001) Emotional Distress Regulation Takes Precedence Over Impulse Control: If You Feel Bad, Do It! *Journal of Personality and Social Psychology,* **80/1,** 53–67.

[133] Treacy,M. and Wiersema,F. (1995) *The discipline of market leaders : choose your customers, narrow your focus, dominate your market* 2, Addison-Wesley, Reading.

[134] Trommsdorff,V. (1993) *Konsumentenverhalten,* 2, Kohlhammer, Stuttgart.

[135] Uncles,M.D., Dowling,G.R., Hammond,K. (2003) Customer Loyalty and Customer Loyalty Programs. *Journal of Consumer Marketing,* **20/July/4,** 294–316.

[136] Van Trijp,H.C.M., Hoyer,W.D. and Inman,J.J. (1996) Why Switch? Product Category-Level Explanations for True Variety-Seeking Behav-

ior. *Journal of Marketing Research,* **33/Aug.**, 281–292.

[137] Wells,W.D. (1986) Three useful ideas. *Richard J. Lutz (Ed.) Advances in Consumer Research,* **13**, 9–11.

[138] West,P.M. (1996) Predicting Preferences: An Examination of Agent Learning. *Journal of Consumer Research,* **23/June,** 68–80.

[139] Westbrook,R.A. and Black,W.C. (1985) A Motivation-Based Shopper Typology. *Journal of Retailing,* **61/1**, 78–103.

[140] White,L. and Yanamandram,V. (2004) Why customers stay: reasons and consequences of inertia in financial services. *Managing Service Quality,* **14/2/3**, 183–194.

[141] Wirth,J. and Chew,P. (2002) The effects of incentives, deal proneness, satisfaction and tie strength on word-of-mouth behaviour. *International Journal of Service Industry Management,* **13/2**, 141–162.

[142] Yeung,C.W.M. and Wyner,R.S. (2004) Affect, Appraisal, and Consumer Judgment. *Journal of Consumer Research,* **31/Sept.** 412–424.

[143] Zajonc,R. (1980) Feeling and Thinking: Preferences Need no Inference. *American Psychologist*, **35/2**, 151–175.

[144] Zeelenberg,M. and Pieters,R. (2004) Beyond valence in customer dissatisfaction: A review and new findings on behavioral responses to regret and disappointment in failed services. *Journal of Business Research,* **57**, 445–455.

Can the Emotion Construct Help Understanding of Contemporary Phenomena in Knowledge Management?

Hans-Peter Liebmann [a,*], Jörg Kraigher-Krainer [b]

[a] University of Graz, Institute of Marketing, Universitätsstraße 15/G3, A-8010 Graz, AUSTRIA
[b] Upper Austria University of Applied Sciences, Campus Steyr, Wehrgrabengasse 1-3, A-4400 Steyr, AUSTRIA

ABSTRACT

Motivation: Although it is a comparably young discipline in the behavioral sciences, the psychology of emotions and moods already affects several other fields of research in management. Yet surprisingly little is known about the influence of emotions on knowledge management. This conceptual contribution attempts to close the gap by offering a literature review of emotions and moods, their influence on cognitive aspects of behavior and the so far implicitly and explicitly discussed role of emotions in the context of knowledge management.

Results: It can be shown that the addressed aspects are closely intertwined and that ignoring emotion psychology in discussions about knowledge management may lead to the danger of not fully understanding the psychological roots of the use, development, and sharing of knowledge. A framework is proposed which could help in bridging the gap.

Contact: hans-peter.liebmann@kfunigraz.ac.at; joerg.kraigher@fh-steyr.at

1 INTRODUCTION

It has been suggested that knowledge is the key production factor of the post capitalistic society (Drucker 1993). In fact, knowledge management meanwhile has become one of the most intensively discussed challenges in management literature (e.g. North 1999). Furthermore, many authors in this field of research address the importance of emotions (e.g. Ackerschott 2001, p. 102, Davenport and Prusak 1999, p. 203).

A systematic comparison between these two fields of research for similarities and interdependencies is still lacking. This is surprising, considering that emotion literature deals with questions like the influence of emotions on (1) perception (e.g. Loewenstein and Lerner 2003), (2) processing styles (e.g. Bless, Bohner and Schwarz 1991), (3) storage of knowledge in memory (e.g. Klauer 2000), or (4), the readiness to communicate about knowledge (Kraigher-Krainer 2005) or to act correspondingly (e.g. Damasio 2001).

Although relevant literature seems to be perfectly aware of the impact of emotions such as anxiety (e.g. Siewers 1999), trust (e.g. Davenport and Prusak 1999), or, commitment (e.g. Schmitz and Zucker 1999), on knowledge, these assumptions are rather more intuitive than theoretically founded.

The aim of chapter two is to find a working definition for the term emotion, its dimensionality and differentiation to similar constructs. Chapter three throws a light on the influence of emotions on cognitions and action tendencies. Chapter four reviews knowledge management literature concerning the discussion and importance of emotions in this field of research, summarized in Table 1. In chapter five we attempt to develop a framework for a better integration of emotion psychology in knowledge management. Finally, chapter six draws conclusions and discusses managerial implications, limitations and future research opportunities.

2 THE EMOTION CONSTRUCT

In our *occidental tradition*, emotions have been broadly viewed as pathologic events for centuries by prominent authors like Nietzsche, Kant, or Freud. Kleinginna and Kleinginna (1981) summarize such approaches under the term *disruptive definitions*, the most prominent advocate being Young (1973), who defines emotions as "… an acute disturbance of the individual as a whole, psychological in origin, involving behavior, conscious experience, and visceral functioning." (p. 367)

In his well recognized contribution, Zajonc (1980) searches for attempts in cognitive psychology to deal with emotions and concludes: "Contemporary cognitive psychology simply ignores affect. The words *affect, attitude, emotion, feeling,* and *sentiment* do not appear in the indexes of any of the major works on cognition [...] nor do these concepts appear in Neiser's (1967) original work that gave rise to the cognitive revolution in experimental psychology." (p. 152) Later works of cognitive psychologists adjust this lack. Anyway, these *cognitive theories of emotions*, most of them derived from *appraisal theory* (Lazarus 1991), conceptualize emotions as the result of cognitions (Elster 1998, Frijda, Manstead and Bem 2000, Mandl and Reiserer 2000).

Meanwhile, a great deal of empirical research contradicts this assumption, which had already been doubted by Zajonc with his notion: "In fact, for most decisions, it is extremely difficult to demonstrate that there has actually been any prior cognitive process whatsoever." (1980, p. 155). We often *decide for X*, simply because we *like X*. Of course, when later asked, we tend to use cognitions to justify the decision. This may be one of the reasons why the term emotion has been redefined within the last two decades (Loewenstein 2000) and why "… recent influential research on emotions highlights both

a. the essential functions served by emotions in coordinating cognition and behavior

b. the detrimental consequences associated with ignoring emotions." (p. 31).

The problem of defining the term *emotion* is best described by authors like Schmidt-Atzert (1996, p. 18) or Otto, Euler and Mandl (2000, p. 11), who point out that the term seems obvious, as long as we are not asked to define it. For the purposes here, we use the following working definition: Emotions (a) give rise to affective experiences, especially feelings of arousal and pleasure/displeasure; (b) generate cognitive processes such as emotionally relevant perceptual effect, appraisals, labelling processes; (c) lead to behavior that is often, but not always, expressive, goal-directed, and adaptive; and (d) have a communication function within the social environment (for similar definitions see Kleinginna and Kleinginna 1981, p. 355; Frijda, Manstead and Bem, 2000, p. 5). By this we seek to take into consideration the complex and interactive effect of emotions, the dominant meaning of activation/disactivation and pleasure/displeasure, the cognition and behavior triggering role of emotions – e.g. strong emotions like flight anxiety resist cognitions - and the communication function of emotions for the subject and his/her environment, which, in turn, can reinforce or subdue them.

Although there is no final consensus about the dimensionality of the construct emotion, in the context lying before two widely accepted dimensions and the resulting classes of emotions seem to properly describe emotional states and their influence on perception and knowledge storage (see Fig. 1). Furthermore, although the discussion about the differences or communalities, respectively, between terms like *emotion, affect, feeling,* or, *mood* is far from being resolved, in our context we only distinguish *emotion* from *mood*, where we talk of mood, if an emotion is

(1) enduring (hours to weeks), (2) not related to a specific entity, and (3) mainly subconscious or of only peripheral consciousness (see e.g. Bagozzi, Gopinath and Nyer, 1999; Gendolla 2000; Gross 1998; Pham 1998; Schwarz and Clore 2003).

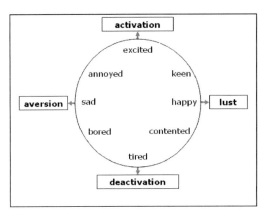

Fig. 1. A two-dimensional model of emotion. Source: Hänze (1998, p. 19); transl. by authors

3 THE INFLUENCE OF EMOTION ON COGNITION AND ACTION

The question is not, whether or not cognitions influence emotions: Of course, they do. However as "... there is much discussion of the effects of cognitions upon emotions, but very little discussion of the effects of emotions upon cognitions ..." (Frijda, Manstead and Bem 2000, p. 5), this paper focuses on the latter aspect. We suppose a relatively dominant role of emotions, as "... some emotion or pattern of emotions is always present and interacting with the perceptual, cognitive, and motor systems ..." (Izard 1991, p. 44).

Unfortunately, we are addressing a scientific problem which is still unresolved (Ciompi 1998, Drucker 1993). However there are some aspects which can be scrutinized in this important and upcoming field of research, at least for the purpose of throwing a light on possible influences of emotions on knowledge development.

As a starting point of this brief and selective review, we choose the *hedonic man assumption* (see Tice, Bratslavsky and Baumeister 2001 for an excellent and recent contribution), which understands man as a being that seeks pleasure and avoids pain. The activities to reach these goals are summarized under the term *mood management* with its two components: (1) *mood maintenance* to keep up pleasant states and (2) *mood repair* to overcome unpleasant ones (Andrade 2005). Mood management can be seen as a fundamental human motivation: "Happiness, or more generally personal welfare, seems to be a central motive spurring one onto action." (Bagozzi, Gopinath and Nyer, 1999, p. 200) Hence, emotions seem to be strong impulses to act (Lazarus 1991) and should therefore have a strong impact on our cognitions and actions.

One cannot always follow his/her immediate impulses, as they might be in conflict with long term targets or might even cause damage in the long run (e.g., Loewenstein 1996; 2000; Loewenstein and Schkade 1999). Hedonic impulses, therefore, have to be controlled and regulated. The better the individual scrutinizes his/her (or other persons) emotions, the higher his or her *emotional intelligence* will be (Salovey and Mayer 1990). In times of emotional distress an argument like *In two years everything will be better* seems less convincing, as the momentary need for mood repair can eclipse any long term perspective.

A second question deals with the way we handle information when influenced by a different mood. The most prominent assumption, the *mood congruence*-assumption says that the perception of information, its processing, storing, and retrieval from memory are congruent with our immediate moods (e.g. Bower 1981; Gendolla 2000). A somewhat contradictory assumption is that the motivation of mood repair can bias the perception in a way that we prefer contents that seem to help to improve our mood (Forgas 2000). Although the circumstances under which information storage and retrieval is triggered by emotions in detail are still unclear (Hänze 1998; Klauer 2000), it seems clear, that "...immediate emotions can systematically bias the interpretation of decision-relevant information ..." (Lowenstein and Lerner, 2003, p. 22).

A third group of researchers concentrate on the question of how emotion and mood influence *processing styles*. The famous work of Isen and her colleagues - starting about three decades ago - largely contributed to establishing the idea that the decider tends to use a more systematic style of problem solving under bad mood, whereas under good mood the style seems to be more generous and creative, but less exact. More recent research (e.g. Bless, Bohner and Schwarz, 1991; Bless et al. 1996; Fiedler and Bless 2000; Schwarz and Clore 2003) adjusts this assumption by pointing out, that people under good mood do not miss information but prefer to use their capacities for creative associations, whereas "... negative mood states should induce a conservative set to adhere to the input data as carefully as possible." (Fiedler and Bless 2000, p. 147, similar Loewenstein 1996). Again, mood repair can additionally influence the decision process, as allocating mental capacity to the suppression of negative emotions reduces the free capacities for problem solution (Ellis, Seibert and Varner 1995; Richard and Gross, 2000).

A fourth field concerning the question how emotions influence cognitions and actions, is the approach that emotions do not only influence information processing, but rather *are* information themselves (e.g. Schwarz and Clore, 1983; 2003), basically represented by the *how do I feel-heuristic*: When people are asked about their satisfaction with life, they tend to be less satisfied on rainy days than on sunny days. They seem to take their momentary – yet irrelevant – feelings as information to judge their overall life situation. Consequently, this effect disappears – at least on rainy days – if people are pointed to the bad weather. Negative feelings can deactivate people and even paralyze them. On the other hand, people in a good mood seem more optimistic, they might project their momentary feelings into the future (Loewenstein 1996). Damasio and Damasio (1994) are of the opinion that without emotion there is no action, as only emotions have the capacity to interrupt cognitions and LeDoux offers a biological explanation: "You have to stop thinking about whatever you were thinking about before the danger occurred and start thinking about the danger you are facing" (LeDoux, 1996, p. 176). Hence, following these informational qualities of emotions, they do no more appear as pathologic events. It also seems misleading, "... to assert that emotions are a »supplemental« principle that »fills the gap« between reflex-like behavior and fully rational action ... the emotion serves as *a functional equivalent for the rational faculties it suspends*." (Elster 1998, p. 60)

To sum up this very compressed literature review, it can be stated that information, knowledge, and emotions are complexly interrelated phenomena, and that discussing knowledge management from a solely cognitive perspective would in fact thwart the access to a holistic view of people and how they deal with knowledge. To further investigate this question, the next chapter offers a review of the impact of emotions from a knowledge management-perspective.

4 EMOTIONS IN KNOWLEDGE MANAGEMENT LITERATURE

While trying to create such a review in an objective manner, we basically face two problems:

(1) The literature offers a variety of emotional phenomena in describing emotion-laden situations but without conceptualizing the involved constructs;

(2) As a result, it turns out to be even more difficult to decide whether emotional, motivational, attentional or other phenomena are addressed by the cited author(s).

However, we decided for this intuitive kind of selection from the literature, although we are aware of the subjective nature of this process. Still, it would not have helped to search simply for key terms like *emotion, affect, feeling,* or *mood*. Hence we picked up all passages which can be categorized as emotional phenomena from the emotion literature's perspective.

General Importance: Some authors in fact suspect, that emotions have some kind of importance, that they are more important than technology or organization (Ackerschott 2001) or that anxiety, envy, or hatred play a role (Grässle 1999). Dammermann-Prieß (1999, p. 282) or Schneider (2001, p. 25) propose emotional learning curves involving phases like euphoria, disillusion, crises, or crashes. Davenport and Prusak (1999) talk of emotional reasons not to use existing knowledge (e.g., lack of confidence in source, lack of reputation of source, pride, stubbornness, risk aversion in companies with little tolerance for mistakes). Güldenberg (2001) contrasts change of organizational thinking from change of organizational feeling and Nonaka and Takeuchi (1995, p. 53) distinguish logical from nonlogical – which can be understood as emotional - processes of »good judgment« or »good sense«.

Trust and Anxiety: A specific emotion that can help to promote the development of knowledge is to give the employees a sense of crisis before a crisis actually can develop (Nonaka and Takeuchi 1995; Davenport and Prusak 1999). The most prominent advocate for this thesis is Edgar H. Schein, who assumes that "…all learning is fundamentally coercive because you either have no choice ... or it is painful to replace something that is already there with some new learning." (Coutu 2002, p. 103). With this approach, Schein taps on a phenomenon that has to do with the above mentioned hedonic principle or mood management: anxiety hampers learning and change, because new behavior could be threatening and we would have to overcome our beloved habits - *learning anxiety* - or in terms of emotion psychology *mood maintenance*. On the other hand we experience *survival anxiety* – an emotion that has to do with the necessity of change to be happy in the long run. Only if survival anxiety exceeds learning anxiety will people's behavior be expected to change. Managers, therefore, have basically two options: to reduce learning anxiety, or to raise survival anxiety. As the latter mood is easier to manipulate ("learn, or …") Schein suggests that most managers tend to prefer this option. This, again, is what Siewers (1999) calls the trap in change management: a company that permanently produces pressure and anxiety will not survive (p. 139f). He sees trust and anxiety as antagonists and proposes to develop an atmosphere of trust and confidence instead of anxiety and control.

Another context in literature where trust is discussed is the readiness of employees to exchange and share knowledge. People do not automatically share their knowledge; several barriers have to be overcome to make knowledge flue (Probst, Raub and Romhardt, 1999). Davenport and Prusak (1999, p. 83) see trust as the first and must important mean in case it meets three conditions: it is visual; it is omnipresent; it is truly lived top-down. Schmitz and Zucker (1999) similarly, see trust and sympathy between people as important factors; following Papmehl (1999) it can be suggested that these people build up trustworthy partnership-groups where mutual win-win-situations can grow. Schneider (2001) points at the problem that, otherwise, each employee tries to maximize his taken-given-ratio which results in a far weaker organisational success than if all members maximize both, giving and taking knowledge, which is not corresponding with the individualistic compensation- and assessment-systems in many companies (North, 1999).

Taken together, there seems to be comparable consensus about trust vs. distrust as an important emotional factor to support vs. hinder knowledge flow. Concerning anxiety, the opinions are quite contrary; reaching from *all learning is coercive* to *permanent anxiety ruins a company*. Later, we will return to the anxiety construct, discussing specific forms of anxiety in organizations.

We-Feeling and Inner Dismissal. Grässle (1999, p. 60) sees the corporate community as one of the most important contexts for defining the individuals personality and his or her respective responsibility. If the community gives the individual a sense of being part of it, this We-Feeling results in strong group coherence, whereas the opposite leads to a sense of being left alone and, as a consequence, to the emotional status of Inner Dismissal. Similarly, North

(1999) suggests, that even very talented people, would not lead to a sufficient group performance, in case this We-Feeling is undermined by rivalry, arrogance, or, authoritarian culture. Authoritarian culture combined with pressure for immediate attention and intimidation cause an emotional atmosphere, where the individual tends to emotionally abandon the organization and its objectives (Schmitz and Zucker, 1999). Especially, if the tasks become increasingly monotonous, the danger of Inner Dismissal and decreasing share of knowledge can be expected (Güldenberg 2001) and might even lead to sabotage and disinformation (Probst, Raub and Romhardt, 1999).

A second discussed aspect, which has to do with We-Feeling, is the *not invented here-syndrome* and taps the ability of organizations to utilize external knowledge, yet in the opposite direction: it tones down knowledge and causes a tendency to reject knowledge simply because this knowledge comes from outside the group (Davenport and Prusak, 1999; North, 1999; Probst, Raub and Romhardt, 1999).

Anxiety of mistakes and exposure. Some discussion can be found on anxiety of exposure resulting from mistakes. People in general seem to experience any action that could harm their image, or their self-image, or that could lead to uncovering their past decisions as weak ones, as extremely threatening. As a result, people tend to resist change, learning, and knowledge. Much of the prominent work of Argyris is dedicated to different forms of such *information pathologies* (e.g. 1997, p. 29) and consequent *defence strategies* (p. 102f) including emotions such as awkwardness, dismay, doubt, cynicism, attack, exploitation, or exhaustion. All these defence strategies dramatically decrease the probability of individuals or organizations to learn and improve. They are subtle strategies to put certain solutions under taboo and can cause grave disturbances of innovation processes (Probst, Raub and Romhardt 1999). Each time, a person has to commit its smaller knowledge learning becomes painful (Papmehl 1999).

Anxiety of losing power. Comparably ample discussion is applied to anxiety of losing power and influence. Intransparency seems to be a proper method of keeping up influence that relies on knowledge lead (Probst, Raub and Romhardt 1999). As knowledge then equals power, it is locked away from others (North 1999). Knowledge can then be used as a weapon to blame others and to produce anxiety (Romhardt 2001), team spirit is replaced by individual competition (Ackerschott 2001). Romhardt (2001) speaks of the *political organisation* with well informed circles that control power and knowledge. Information becomes a tool of personal profit which in turn promotes distrust and disinformation. The organisation becomes helpless by and by. Following Güldenberg (2001) this kind of behaviour not only results from the tendency to increase power by passing on information selectively or distortedly, it has also to do with the anxiety of getting useless within the organisation, if the personal knowledge lead become diminished or absorbed. Papmehl (1999) adds as a motive that only 80 percent of knowledge may be passed over to ensure that a colleague's maximum performance can only stand at 80 percent.

Although this aspect of emotional influence on knowledge growth gets much attention in literature it remains unclear whether it should be understood as an emotional phenomenon, or rather as rational calculation: loss of power means loss of opportunities and, finally, loss of money.

Summarizing the Review. Regarding the contents of the literature review it can be stated that emotions are widely discussed in knowledge management literature but without offering information that could help understanding the relationships between the different terms and constructs. Considering the fact that emotions are ignored in other fields of behavioral sciences, knowledge management seems at least better developed than – for instance – cognitive psychology. The identified emotional aspects are listed in an overview in table 1.

Table 1. An overview of the identified emotions that can promote or hinder knowledge growth.

emotions that promote knowledge growth	emotions that hinder knowledge growth
• Sense of Crisis	• Lack of Trust
• Anxiety	• Stubbornness, Pride, Doubt, Cynicism
• Trust	
• Sympathy	• Risk Aversion, Anxiety of Loosing Power, Influence, or Prestige
• We-Feeling	
	• Rivalry, or Arrogance
	• Not Invented Here-Syndrome

Regarding the methodological aspects of the literature review it has to be noticed that the role of emotions and moods in knowledge management still seems to be in its infancy. Three aspects deserve special attention:

(1) A contribution with a conceptual framework cannot be found;
(2) Conceptual or operational construct definitions are lacking;
(3) Empirical support is neither strived for, nor available.

5 A FRAMEWORK ON EMOTION AND KNOWLEDGE

Regarding the aim of this article, we now propose a theoretical framework, to stimulate a more detailed and empirically founded discussion about the influence of emotions on cognition, knowledge development and vice versa. The reviews of the two addressed fields of research are combined and depicted in Fig. 2. Next we give brief reasoning for our assumptions. As a general model underlying the framework we use the ECID-Model (Kraigher-Krainer 2005), which has been developed in the field of consumer research but is proposed here to be applicable to other managerial decision processes as well.

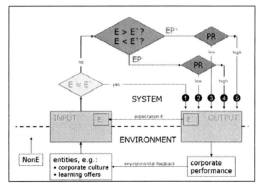

Fig. 2. A framework concerning the interdependencies between emotion and cognition in knowledge management. Description of constructs see text.

The framework shows a cybernetic loop with its standard elements *system* above the dashed line (with connection to environment via *input* and *output*) and *environment* below the line. System and environment have to be strictly separated: Either one is part of the environment, then he cannot look into the system – basically man's brain - or he is the system, then he depends on what his or her input offers. For illustrative reasons, we look at both perspectives at the same time now and describe the elements of Figure 2 in clockwise order, starting with the system environment.

Entities and *Non-Entities (NonE)*. The comparably tiny amount of objects which are identified by the input system is termed an *entity*, be it a person, a thing, a sound, an organization etc., the bulk of remaining entities is called *NonE*. From the system's perspective, NonE do not exist – independently from what another person observes. Here we concentrate on three such entities: Corporate Culture, Learning Offers, and Corporate Performance, all of them being subjectively interpreted by the system. *Corporate Culture* addresses aspects like trustworthiness of the organization. For example and as mentioned above, trust has to be visual, omnipresent, and lived top-down, otherwise it is a NonE. It encloses not only, how one personally is treated, but also, how colleagues are treated, We-feeling, tolerance of mistakes, or, how much competition, rivalry, or arrogance exists between the members of the organization. Whether or not *learning offers* (colleagues, consultants, customers, seminars, databases, daily problems ...) are perceived, depends on the constructs described below. As *corporate performance* is conceptualized as part of the input system (see *sense of crisis* mentioned above) as well as the result of behavior it is depicted separately.

Input. The logic of the input-system follows the expectation-disconfirmation-paradigm (e.g. Oliver 1980). An expectation can be described as a thesis of the system about what will happen next. Entities other than our expectations are not perceived, whereas entities of the same kind but significantly more or less than expected cause an emotional perturbation (pleasant surprise or unpleasant disappointment). Of course, we have a certain set of entities (e.g. dangerous ones), which are always present. Entities that roughly fit our expectations lead to habitual behavior which is defined as behaving the way the system behaved last time, this entity occurred (see E≈E`, and No. 1 – output in Fig. 2). By this, we not only address Schein's opinion that negative emotions can help overcome our beloved habits (see EP^-) but as figure 2 shows, we conceptualize a second learning-opportunity, caused by positive emotions (EP^+) or positive deviation of the input from expectation. We also point at the phenomenon of subjective perception: people in a good mood may tend to avoid entities that could threaten it (mood maintenance), people in a bad mood may seek for entities to light up their lives (mood repair). Additionally, perception might be biased by the tendency of mood-congruent contents. Furthermore, we do not regard habits as something that is always undesirable. Habits – although widely seen as something negative – could turn out to be our most powerful system without which no learning in little pieces can take place - see for instance Druckers (1993) *techne*, Weizsäckers (1974) *pragmatic informa-*

tion, or Kandel and Hawkins´ (1994) and Schneiders (1996) *implicit learning*. Finally it is worth noting, that small deviations from our expectations are *assimilated* (Glasersfeld 1991; 1997) and that the tolerance for such deviations can grow by adapting our expectations in case we experience little motivation or ability to change things.

Emotional Perturbation (EP). It is important to note that so far we have only described processes that do not require attention or cognition. Our psychic system is engaged only if an entity significantly deviates from expectation. A positive and surprising deviation is called EP^+ in Fig. 1, a negative and disappointing deviation is called EP^-. Without such a primary emotion no cognition can take place. This means, we think that emotions prime cognitions. But as the framework is cybernetic, this is only true for the initializing event of thoughts. From the second loop on it is also true that cognitions cause emotions and that the ongoing cycles produce complex patterns of emotions and thoughts. Furthermore, only after such a primary emotional perturbation, can a learning process in the sense of behavioral change take place. Following Damasio we additionally suppose that any decision is brought to an end by an emotional, final *yes*, before we start acting. In other words: Any cognition is stirred up and finalized by an emotion, which emphasizes the importance of emotions in understanding cognitions, decisions and behavioral changes. Ongoing EP^--series or EP^+-series define our personal mood and additionally serve as information. Remember that we tend to be optimistic and think more creatively if in a positive mood, whereas we adhere to the input data and tend to be skeptical and conservative, if negative mood is present. This makes obvious that producing stress and control can in fact lead to change, but probably restricted by the kind of thinking and problem solving mentioned. Furthermore, we suppose, that under good mood we put significantly more energy into our efforts. These two emotional states enriched with a motivational aspect can be termed intrinsic vs. extrinsic motivation (e.g. Ryan and Deci 2000). Intrinsic motivation is the motivation to do something because the activity itself is pleasant. Extrinsic motivation is the motivation to do something although this activity is not pleasant per se, but helps in achieving a goal or avoiding pain (compare to the hedonic man-assumption above). If we experience restrictions in our capacity (cognitive, financial, time, etc.), even an intrinsic motivation alters to an extrinsic motivation determined by EP⁻-series of disappointment.

Perceived Risk (PR in Fig. 2). Like most authors in the field of knowledge management point out, people prevent from sharing their knowledge for fear of losing something like image, competence, power, influence, control, importance or the like. We also mentioned that it remains unclear whether this phenomenon should be seen as an emotional one (anxiety), or as a cognitive one (loosing opportunities and finally loosing money). In ECID-model this aspect is represented by the construct *perceived risk*, defined as the subjectively perceived probability of unexpected loss of resources and is clearly understood as a cognitive phenomenon. Resources are entities that are suitable for reaching pleasant states, entail things like food, money, attractiveness, or power, and differ from one person to another, from one organization to another, from one culture to another; they are socially constructed (Berger and Luckmann 1977). It is obvious that perceived risk is closely related to corporate culture (see Fig. 2; e.g., values, visions, missions, trust, rivalry, consequences of mistakes, measurement systems). Now the interesting thing is how the individual defines this *system*, which threatens to cause loss of something. We hypothesize that in companies with predominantly positive moods accompanied by a strong We-feeling, people tend to see corporate or group loss similar to personal loss (see Spencer-Brown 1997 for the explanatory power of drawing a distinction), whereas in companies with predominantly negative emotions (produced by sense of crisis, control, authority, cynicism, or rivalry) the organizational task thrusts into the background and loss is more directly related to personal objectives like career, power, or income.

Processing Styles (No. 2 – 5 in Fig. 2). In ECID-model it is proposed that the two different emotional states (EP$^+$ vs. EP$^-$) combined with the two different cognitive states (PR^{high} vs. PR^{low}) result in four processing styles entailing different perception preferences, problem solving heuristics, and action tendencies. Processing styles 2 and 3 have in common that little risk is perceived – although under different mood. The aim is to fix a problem that has been caused by the perturbation and afterwards to continue with the routines. More interesting from the knowledge management perspective are processing styles 4 and 5. In these cases the employee perceives a certain threat, in processing style 4 predominately personal risk, and in style 5 personal and organizational risk. The employee now has to reduce risk, before (s)he can solve the problem. Scrutinizing information and knowledge from within or outside the company is not the only risk reduction strategy but it is an important one. It can be assumed that under positive mood (path 5) the employee will try to reduce both corporate risk and personal risk, and (s)he will combine different methods (like collecting information, consulting colleagues, claiming for guarantees or modified terms of delivery) in a creative and innovative way, possibly using free capacities for exceeding the question at hand to utilize from the activity for future purposes. However, negative mood states should induce a risk reduction strategy more likely focused on personal risk, applying a conservative set to adhere exactly to the input data. As problem solving needs mental capacity and as under negative mood part of these capacities are used for mood repair, the free capacities for the problem are restricted.

Another interesting aspect of the studies in Kraigher-Krainer (2005) is the communication behavior of consumers in path 4 vs. 5: The latter do not only significantly utilize more available information of all kinds, they also actively communicate their expertise to others. Anyway, consumers in path 4 show a significant preference for interpersonal communication. They are opinion seekers and tend to rely on others recommendations while seeking a *good* decision. Their dilemma is: solving the problem superficially causes high personal risk; carefully dealing with the problem prolongs the unpleasant state. For them, relying on the personal help and knowledge of experts and others who are trustworthy is a good compromise between a quick and a personally elaborated solution. In case these results can be transferred to the field of knowledge management, the disposal of at least some employees who work under positive mood-conditions seems to be crucial for the growth and distribution of knowledge within the company.

Corporate Performance. Finally we assume that the different paths of activity within a company, and its respective combinations, trigger corporate performance and this, in turn, triggers corporate culture and learning offers, e.g. affecting *senses* of crisis, as well as *real crises*.

6 CONCLUSIONS

If any learning process is primed by and made up by an emotion, we don't think that we go too far if we say: Knowledge management is primarily a task of emotion management. This emotion management entails the management of surprise and disappointment, which is proposed to be the starting point of any change in thinking and behaving – for a similar approach see Ansoff's (1976) approach of *Managing Surprise and Discontinuity*. It also entails a corporate culture which allows employees a final, emotional *yes*, to make sure that decision processes come to an end one day. It even entails the management of routines, as by far not every task should be a matter of permanent rethinking and tolerances have to be defined, that allow employees to assimilate deviations from their expectations.

From an emotional point of view we speculate that there are basically three kinds of prototype organizations:

(1) The $E{\approx}E`$-organization, which focuses on routinized processes, dislikes all kinds of deviations from these routines and hardly learns anything;

(2) the EP^--organization, which likes to disappoint and control, learns in competing atmosphere by avoiding mistakes within these controls, preferably perceives traps, mistakes of colleagues and reasons why things do not work, is specialized on *doing the things right*, focuses on the loss of personal resources like power, salary or company car, suffers from restricted capacities for problem solving, as part of these capacities are used for mood repair and inner competition and lacks experts that are willing and able to distribute their expertise to others;

(3) the EP^+-organization, which likes surprise and challenge, learns within a positive atmosphere and we-feeling, tends to be creative and innovative and to push the limits, preferably perceives opportunities, optimistic information and approach-targets, is specialized on why things **do** work and on *doing the right things*, considers personal **and** organizational risk, has more free capacity for problem solutions and is in the possession of intrinsically motivated opinion leaders, who like to distribute their expertise within the organization.

As mentioned above, the first and most important responsibility of managers to transfer a static into a dynamic organization seems to be the management of surprise, disappointment and routine. The ratio of these three events within the organization defines whether it will rather become an $E{\approx}E`$-, EP^--, or, EP^+-organization. Furthermore, perceived risk as the cognitive aspect of knowledge management has to be handled. This task entails (1) the support with risk reduction tools, e.g. information of all kinds; (2) the promotion of corporate risk and the reduction of personal risk as the latter grows automatically whereas the former tends to thrust into the background, esp. in E≈E'-, and EP'-organizations; (3) a ranking of the – basically unlimited – aspects of corporate risks; e.g. by transparent value systems, or score cards; (4) encouraging employees to finalize the risk reduction process by an emotional *yes*.

Considering the momentary state of work in this domain our conclusions are predominantly of a speculative nature. Empirical and even theoretical support for our theses is still poor. However one has to start somewhere and the presented conceptualization could be an assisted takeoff for stimulating more research on the impact of emotions on knowledge management. Such empirical research could include an empirical foundation of the conceptualization per se. The question, whether a pure form – and which – of the three prototypes is more successful is of further interest. Or is it rather a certain mix and if, which? – ideally related to measurable aspects like profit. Moreover the company's environment (e.g., the speed of change within an industry), the core business of the company (e.g., entertainment vs. administration) or even the core business of departments (take, for instance, the sales department contrasted to the technical crew of an airport) might have an influence on the decision as to how emotions could be utilized to improve company's knowledge performance in the short run, and its overall performance, in the long run.

7 REFERENCE

[1] Ackerschott,H. (2001) *Wissensmanagement für Marketing und Vertrieb. Kompetenz steigern und Märkte erobern.* Gabler, Wiesbaden.

[2] Andrade,E.B. (2005) Behavioral Consequences of Affect: Combining Evaluative and Regulatory Mechanisms. *Journal of Consumer Research*, **32/Dec.**, 355–362.

[3] Ansoff,H.I. (1976) Managing Surprise and Discontinuity – Strategic Response to Weak Signals. *Schmalenbachs Zeitschrift für betriebswirtschaftliche Forschung*, **28/3**, 129–130.

[4] Argyris,C. (1997) *Wissen in Aktion. Eine Fallstudie zur lernenden Organisation.* Klett-Cotta, Stuttgart.

[5] Bagozzi,R.P., Gopinath,M. and Nyer,P.U. (1999) The Role of Emotions in Marketing. *Journal of the Academy of Marketing Science*, **27/2**, 184–206.

[6] Berger,P.L. and Luckmann,T. (1977) *Die gesellschaftliche Konstruktion der Wirklichkeit. Eine Theorie der Wissenssoziologie.* Fischer, Frankfurt/Main.

[7] Bless,H., Bohner,G. and Schwarz,N. (1991) Gut gelaunt und leicht beeinflussbar? Stimmungseinflüsse auf die Verarbeitung persuasiver Kommunikation. *Psychologische Rundschau*, **43**, 1–17.

[8] Bless,H., Clore,G., Schwarz,N., Golisano,V., Rabe,C. and Wölk,M. (1996) Mood and the use of scripts: Does happy mood make people really mindless? *Journal of Personality and Social Psychology*, **71**, 665–679.

[9] Bower,G. (1981) Mood and Memory. *American Psychologist*, **36**, 129–148.

[10] Ciompi,L. (1998) *Affektlogik. Über die Struktur der Psyche und ihre Entwicklung.* Klett-Cotta, Stuttgart.

[11] Coutu,D.L. (2002) The Anxiety of Learning. *Harvard Business Review*, **3**, 100–106.

[12] Damasio,A.R. (2001) *Descartes' Irrtum. Fühlen, Denken und das menschliche Gehirn.* DTV, München.

[13] Damasio,A. and Damasio,H. (1994) *Sprache und Gehirn.* In: Singer,W. (Ed.) Gehirn und Bewusstsein. Spektrum, Heidelberg, 58–66.

[14] Dammermann-Prieß,G. (1999) *Team-Lernen in der Führungskräfteentwicklung.* In: Papmehl,A. and Siewers, R. (Eds.) Wissen im Wandel. Die lernende Organisation im 21. Jahrhundert. Ueberreuter, Wien, 264–287.

[15] Davenport,T.H. and Prusak,L. (1999) *Wenn Ihr Unternehmen wüsste, was es alles weiß ... Das Praxisbuch zum Wissensmanagement.* Moderne Industrie, Landsberg/Lech.

[16] Drucker,P.F. (1993) *Die postkapitalistische Gesellschaft.* Econ, Düsseldorf.

[17] Ellis,H.C., Seibert,P.S. and Varner,L.J. (1995) Emotion and Memory: Effects of Mood States on Immediate and Unexpected Delay Recall. *Journal of Social Behavior and Personality*, **10/2**, 349–362.

[18] Elster,J. (1998) Emotions and Economic Theory. *Journal of Economic Literature*, **36/1**, 47–74.

[19] Fiedler,K. and Bless,H. (2000) *The formation of beliefs at the interface of affective and cognitive processes.* In: Frijda,N.H., Manstead,S.R. and

Bem,S. (Eds.) *Emotions and Beliefs. How Feelings Influence Thoughts.* Cambridge University Press, Cambridge, 144–170.

[20] Forgas,J.P. (2000) *Feeling is Believing? The role of processing strategies in mediating affective influences on beliefs.* In: Frijda,N.H., Manstead,S.R. and Bem,S. (Eds.) *Emotions and Beliefs. How Feelings Influence Thoughts.* Cambridge University Press, Cambridge, 108–143.

[21] Frijda,N.H., Manstead,S.R. and Bem,S. (2000) *The influence of emotions on beliefs.* In: Frijda,N.H., Manstead,S.R. and Bem,S. (Eds.) *Emotions and Beliefs. How Feelings Influence Thoughts.* Cambridge University Press, Cambridge, 1–9.

[22] Gendolla,G.H.E. (2000) On the impact of Mood on Behavior: An Integrative Theory and a Review. *Review of General Psychology,* **4/4**, 378–408.

[23] Glasersfeld,E.v. (1991) *Einführung in den radikalen Konstruktivismus.* In: Watzlawick,P. (Ed.) *Die erfundene Wirklichkeit. Wie wissen wir, was wir zu wissen glauben. Beiträge zum Konstruktivismus.* Piper, München, 16–38.

[24] Glasersfeld,E.v. (1997) *Radikaler Konstruktivismus. Ideen, Ergebnisse, Probleme.* Suhrkamp, Frankfurt/Main.

[25] Grässle,A.A. (1999) *Von der lernenden Organisation über Netzwerke zur „Corporate Community".* In: Papmehl,A. and Siewers,R. (Eds.) *Wissen im Wandel. Die lernende Organisation im 21. Jahrhundert.* Ueberreuter, Wien, 36–65.

[26] Gross,J.J. (1998) The Emerging Field of Emotion Regulation: An Integrative Review. *Review of General Psychology,* **2/3**, 271–299.

[27] Güldenberg,S. (2001) *Wissensmanagement und Wissenscontrolling in lernenden Organisationen. Ein systemtheoretischer Ansatz.* Deutscher Universitätsverlag, Wiesbaden.

[28] Hänze,M. (1998) *Denken und Gefühl. Wechselwirkung von Emotion und Kognition im Unterricht.* Luchterhand, Neuwied.

[29] Izard,C.E. (1991) *The Psychology of Emotions.* Plenum Press, New York.

[30] Kandel,E. R. and Hawkings,R.D. (1994) *Molekulare Grundlagen des Lernens.* In: Singer,W. (Ed.) *Gehirn und Bewußtsein.* Spektrum, Heidelberg, 114–124.

[31] Klauer,K.C. (2000) *Gedächtnis und Emotion.* In: Otto,J.H., Euler,H.A. and Mandl,H. (Eds.) *Emotionspsychologie.* Beltz, Weinheim, 315–324.

[32] Kleinginna,A.M. and Kleinginna,P.R. (1981) A Categorized List of Emotion Definitions, with Suggestions for a Consensual Definition. *Motivation and Emotion,* **5/4**, 345–379.

[33] Kraigher-Krainer,J. (2005) *Das ECID-Modell. Konzeptualisierung, Operationalisierung, empirische Prüfung und strategische Implikationen.* Habilitationsschrift, Graz.

[34] Lazarus,R.S. (1991) *Emotion and Adaption.* Oxford University Press, New York.

[35] LeDoux,J. (1996) *The emotional brain. The mysterious underpinnings of emotional life.* Touchstone, New York.

[36] Loewenstein,G. (1996) Out of Control: Visceral Influences on Behavior. *Organizational Behavior and Human Decision Processes,* **65/3/March**, 272–292.

[37] Loewenstein,G. (2000) Emotions in Economic Theory and Economic Behavior. *AEA Papers and Proceedings,* **90/2**, 426–432.

[38] Loewenstein,G. and Lerner,J.S. (2003) *The Role of Affect in Decision Making,* in Davidson,R.J. et al. (Eds.) *Handbook of Affective Science.* Oxford University Press, Oxford, 619–642.

[39] Loewenstein,G. and Schkade,D. (1999) *Wouldn't It Be Nice? Predicting Future Feelings.* In: Kahneman,D., Diener,E. and Schwarz N. (Eds.) *Wellbeing: The foundations of hedonic psychology.* Russell Sage, New York, 85–108.

[40] Mandl,H. and Reiserer,M. (2000) *Kognitionstheoretische Ansätze.* In: Otto,J.H., Euler,H.A. and Mandl,H. (Eds.) *Emotionspsychologie.* Beltz, Weinheim, 95–105.

[41] Nonaka,I. and Takeuchi,H. (1995) *The Knowledge-Creating Company. How Japanese Companies Create the Dynamics of Innovation.* Oxford University Press, New York.

[42] North,K. (1999) *Wissensorientierte Unternehmensführung. Wertschöpfung durch Wissen.* Gabler, Wiesbaden.

[43] Oliver,R.L. (1980) A Cognitive Model of the Antecedents and Consequences of Satisfaction Decisions. *Journal of Marketing Research,* **17/Nov.**, 460–469.

[44] Papmehl,A. (1999) *Wer lernt, ist dumm! Stolpersteine auf dem Weg zur Wissensorganisation.* In: Papmehl,A. and Siewers,R. (Eds.) *Wissen im Wandel. Die lernende Organisation im 21. Jahrhundert.* Ueberreuter, Wien, 228–247.

[45] Pham,M.T. (1998) Representativeness, Relevance, and the Use of Feelings in Decision Making.

Journal of Consumer Research, **25/Sept.**, 144–159.

[46] Probst,G., Raub,S. and Romhardt,K. (1999) *Wissen managen. Wie Unternehmen ihre wertvollste Ressource optimal nutzen.* FAZ-Verlag, Frankfurt/Main.

[47] Richards,J. and Gross,J. (2000) Emotion Regulation and Memory. The Cognitive Costs of Keeping One´s Cool. *Journal of Personality and Social Psychology*, **79/3**, 410–424.

[48] Romhardt,K. (2001) *Wissen ist machbar. 50 Basics für einen klaren Kopf.* Econ, München.

[49] Ryan,R.M. and Deci,E.L. (2000) Intrinsic and Extrinsic Motivations: Classic Definitions and New Directions. *Contemporary Educational Psychology*, **25/1**, 54–67.

[50] Salovey,P. and Mayer,J.D. (1990) Emotional Intelligence. Imagination, Cognition and Personality, **9/3**, 185–211.

[51] Schmidt-Atzert,L. (1996) *Lehrbuch der Emotionspsychologie.* Kohlhammer, Stuttgart.

[52] Schmitz,C. and Zucker,B. (1999) *Wissen managen? Wissen entwickeln! In: Papmehl,A. and Siewers,R. (Eds.) Wissen im Wandel. Die lernende Organisation im 21. Jahrhundert.* Ueberreuter, Wien, 178–203.

[53] Schneider,U. (1996) *Management in der wissensbasierten Unternehmung. In: Schneider,U. (Ed.), Wissensmanagement.* FAZ-Verlag, Frankfurt/Main, 13–48.

[54] Schneider,U. (2001) *Die 7 Todsünden im Wissensmanagement. Kardinaltugenden für die Wissensökonomie.* FAZ-Buch, Frankfurt/Main.

[55] Schwarz,N. and Clore,G.L. (1983) Mood, Misattribution, and Judgments of Well-Being: Informative and Directive Functions of Affective States. *Journal of Personality and Social Psychology*, **45/3**, 513–523.

[56] Schwarz,N. and Clore,G.L. (2003) Mood as Information: 20 Years later. *Psychological Inquiry*, **14/3&4**, 296–303.

[57] Siewers,R. (1999) *Über den Schatten springen - vertrauen, wagen und lernen. In: Papmehl,A. and Siewers,R. (Eds.) Wissen im Wandel. Die lernende Organisation im 21. Jahrhundert.* Ueberreuter, Wien, 135–176.

[58] Spencer-Brown,G. (1997) *Laws of Form.* Bohmeier, Lübeck.

[59] Tice,D.M., Bratslavsky,E. and Baumeister,R.F. (2001) Emotional Distress Regulation Takes Precedence Over Impulse Control: If You Feel Bad, Do It! *Journal of Personality and Social Psychology*, **80/1**, 53–67.

[60] Weizsäcker,E.v. (1974) *Erstmaligkeit und Bestätigung als Komponenten der pragmatischen Information, In: Weizsäcker,E.v. (Ed.) Offene Systeme I.* Klett, Stuttgart, 82–113.

[61] Young,P.T. (1973) *Feeling and emotion. In: Wolman,B.B. (Ed.) Handbook of general psychology.* Englewood Cliffs, New Jersey.

[62] Zajonc,R. (1980) Feeling and Thinking: Preferences Need no Inference. *American Psychologist*, **35/2**, 151–175.

Football and Mood in Italian Stock Exchange

Claudio Boido[a], Antonio Fasano[b]

[a] Università di Siena, Dipartimento di Studi Aziendali e Sociali,
P.zza S. Francesco, 7 – 53100 Siena, ITALY
[b] Università di Salerno, Dipartimento di Studi e Ricerche Aziendali,
Via Ponte don Melillo – 84084 Fisciano (SA), ITALY

ABSTRACT

Motivation: A number of papers have shown that human behaviour is often inconsistent with the type of rationality that has existed traditionally in finance and economics.[1] People ask whether the behaviour of investors irrationally could impact on asset prices. Often in the absence of new information psychologists show the presence of another source of irrationality: mood.

Our past working papers [2] have regarded the behaviour of individual investor in different situations and our empirical tests have shown that sometimes the irrationally could impact on asset prices.

Our interest is centred on soccer since many situations show that soccer matches bring us laughter and tears, bliss and pain, so they could change the investor's mood.

This study is concerned with the literature on asset pricing anomalies and the goal is to verify whether soccer results have a sufficiently large impact on mood to justify the reaction of prices. We probe Italian market by analysing three Italian football teams: Rome, Lazio, Juventus, which have decided to quote at the beginning of this millennium. We will try to demonstrate if there is a link between mood and stock returns according to the results and the special events concerning the Italian football team.

Results: Soccer data, relative to Italian teams, shows that the average prices/returns immediately after wins are higher than average prices/returns after unsuccessful matches. We examined also the impact of lost and draw matches, to check for possible inconsistent results, and it also appears, on this regard, that Italian investors dislike matches ending in ties.

Contacts: boido@unisi.it, afasano@unisa.it

1 INTRODUCTION

The year 2006 will be remembered as an unforgettable year for Italian soccer supporters for two special events: 1) the victory in the Football World Championship in Germany; 2) the football scandal that has caused the relegation in first

---

[1] Throughout this paper we use the concept of rationality as in the general Efficient Market Theory. We assume that an investor is rational when his or her investment decisions are based on a mathematical process which maximizes expected returns while minimizing expected risks. Given this, "irrationality" is the consequence of not being rational. This is just a convention since there may be investment drivers which are not consistent with this tenet, but that can be considered rational as well, e.g. ethical drivers present in the case of social investing.

[2] Boido, C., Fasano, A., 2006, "Small Caps vs. Big Caps: a comparison based on the impact of market news.", 26th International Symposium on Forecasting, Santander, 11-14 June, Working paper.
Boido, C., Fasano, A., 2006, "News, Prices, Volumes: The Behaviour of small cap investors on Italian Stock Exchange", 30th Anniversary of The Journal of Banking and Finance Conference, Peking University, Beijing, China, 6-8 June . Working paper.
Boido, C., Fasano, A., 2005, "Calendar Anomalies:Daylights Savings Effect" The ICFAI Journal of Behavioural Finance vol II n. 4 december pag 7-24

division of the most representative Italian football club, Juventus.

Many people remember the enthusiastic celebration in Rome (Circo Massimo) with the Italian National Football team and a big crowd of crazy supporters and we think that football could influence the mood of each individual. It is interesting to notice that Italian politicians and the Italian minister of the Economy maintain that the victory in the Football World Championship in Germany could help to improve the level of GDP.

Currently three Italian football teams are quoted in the Italian Stock Exchange (Lazio, Rome and Juventus following the order of entry in Italian market) and after the quotation many economists have debated whatever it was convenient for stockholders to keep the shares, since the value is too subject to the sports results. In fact when in May 2006 the football scandal burst, the value of Juventus and Lazio stocks went down during the trades. The economic damage assessed is high, because the clubs risk losing the most significant gains linked to the television rights (i.e. Sky Television contract).

It is clear that there is a strong connection between the sport results and the reaction of the human behaviour and this is well documented in the literature as we will show in the course of this paper. We will investigate the stock market reaction to the outcomes of soccer competitions to demonstrate that the sports results could have a dramatic effect on mood and that the decline of a football team could have a relevant effect on the economy.

Many existing papers have documented a relation between mood and stock returns. In some case these studies tend to transform interesting anomalies in empirical regularities that could be overworked by traders to take profit. Academics and practitioners analysed equity returns, trying to link anomalous returns with a recurring period of time, this bore a new research area: the so called calendar anomalies, where we should mean as an anomaly any such event that could not be explained using the efficient market theory, or any other ordinary theory prevailing in finance literature. Once an anomaly as been identified, its effect tend to mitigate, since investors understand it and embed it in security prices, these are adjusted on the base of this now public information. Coming to specific anomalies, the most common ones are those concerning given year phases.

The continued presence of various anomalies has provoked ongoing debate related to the possible biases of the sample used, particularly regarding the frequency of the historical series and the selection of shares. Those who assert the theory of market efficiency indicate that such anomalies should disappear, although they may still persist if they are not perceived by traders or if arbitrage is expensive and profit is inadequate.

Behavioural finance [3] is a theory based on concepts and models developed within the realm of cognitive psychology, and asserts that psychological forces prevail upon market operators who should act rationally. Behavioural finance theorists maintain that behavioural characteristics might explain over- and under-reactions observed in historical price series.

Behavioural finance may be applied to classify a variety of irrational reactions displayed by

---

[3] Kahneman, D., Tversky, A., 1979, "Prospect theory: an analysis of decision under risk", Econometrica, 46, 171-185.
Fischhoff B., Slovic P., Lichtenstein S., 1977, "Knowing with certainty: the appropriateness of extreme confidence", Journal of experimental Psicology: Human perception and performance, 3, 552-564.
Gervais, S., Odean, T., 2001, "Learning to be overconfident", Review of Financial Studies, 14, 1-27.
Weinstein, N., 1980, "Unrealistic Optimism about Future Life Events", Journal Of personality and Social Psychology, 39, 806-820.
Buehler, R., Grifin, D., Ross, M., 1994, "Exploring the planning fallacy: Why people underestimate their task completion times", Journal of Personality and Social Psychology, 67, 366-381.
Kahneman, D., Tverski, A., 1974, "Judgement Under uncertainty: heuristics and biases", Science, 185, 1124-1131.
Edward, W., 1968, "Conservatism in human information processing", in Kleinmutz, B., ed., Formal Representation of Human Judgement, New York, John Wiley and Sons.

individual investors. It's possible to catalogue different mistaken behaviours by individual investors:

1. some investors choose a group of shares and become overconfindent in their valuation, without considering other shares excluded in the first selection.
2. Some investors are inclined to maintain in portfolio the losing shares and to sell quickly the winning stocks. In fact they think that the shares will rally their fair value without considering that the first choice could be mistaken.
3. Many investors are inclined to change the portfolio excessively, so they tend to give importance to more recent information, therefore they buy the interested share and sell it when another share is performing better than the one they have in their portfolio.
4. Others think that it is profitable to buy a share of a good company, but this strategy does not imply that the share will reflect its intrinsic value.

2 RELATED LITERATURE

Some authors[4] believe that the football results are a measure of mood. In fact sports results document that fans show a positive or negative reaction in relation to their team performance (see the impact of the mood of Italian team supporters after the great victory against French football team). In every case sports result causes a strong impact on optimism or pessimism of individual investor higher than other variables such as sunlight and temperature, daylight savings. The magnitude of football effect is compared to other effects without offering more profit for the stockholders of football companies.[5]

Some researchers[6] maintain that football results have a strong effect on mood: if you compare hearts attacks during the World Championship (1998 the day when England lost to Argentina) it is possible to notice an increase of heart attacks more than 25%. This situation happens not only in football, but also in other different sports, so we can affirm that sporting events have a large impact on human behaviour. So we say that football is an important part of many people's lives. We can notice that most of people in the world follow football matches (as an example the last final between Italy and France in Berlin had more than 1 billion viewers compared to Super Bowl, Wimbledon or Royal Ascot) this clearly shows how large is football audience.

Many researchers [7] have shown that fans are subject to an allegiance bias – the rendering of biased predictions by individuals who are psychologically invested in expected outcome. If the fans believe that their team will win we can expect a greater effect after the losses than after wins. We can notice that biased belief will imply the existence of predictability in stock returns.

We can notice the football effect, not only on Monday morning but in different days of the week. In fact it is possible to measure the effect after the International Cup matches (Champions League, Uefa) so we can observe the difference linked to the international results.

An other event that could create strong variation is held by the market players, when the players are bought or sold to improve the competitive strength of each team. This situation causes a

---

[4] Edmans, A., Garcia, D., Norli, O., 2005, "Football and stock returns", Working Paper, www.ssrn.com

[5] Hirshleifer, D., Shumway, T., 2003, "Good day sunshine: stock returns and the weather", Journal of Finance, 58 (3) p. 1009, 1032 show that even if the weather change, the average daily return increases by only 9 basis points.

[6] Carroll, D., Ebrahim, S., Tilling, K., Macleod, J., Smith, G., 2002, "Admission from myocardial infarction and World Cup football: database survey", British Medical Journal, 325 p 1439-1442

[7] Markman, K., D., Hirt, E., R., 2002, Social prediction and the allegiance bias", Social Cognitive, n.20, pp. 58-86.
Wann, D., Melcnick, M., Russel, G., Pease, D., 2001, "Sport fans: the psychology and social impact of spectators, Routledge

strong reaction in the mood of the fans and also of the investors, because the quality of the football team could improve or worsen and also the value of the stock could change.

It is reasonable to argue that the effect could be stronger for countries where soccer is followed by a large percentage of the population (for example it is quite simple to individuate the European countries more affected by soccer: Italy, England, Spain, Germany, France where the soccer supporters follow on newspaper, television and other media each item of news concerning their football team). For example In Italy, during the football scandal many television and newspaper reserved the first entry to football instead of Italian economics or politics.

It is interesting to observe the Deloitte report[8] where they say that five country cover 79% of European football matches. Each country produces much higher income than any other countries in Europe. If you compare other countries in the world you can notice that in the top five countries there are leading newspapers exclusively dedicated to sports and particularly to football. The other remarkable geographical group includes South America (Brazil, Argentina and Mexico over all) where the media are interested in football even if the quality of life is not high, and many people pay more for football matches compared to other consumptions.

In the last 20 years interest in football has increased in the economic field; in fact historically football is always considered a working class game and investors did not believe they could get huge gains in this sector.

Recently football club have attracted corporate clients through hospitality, advertising, merchandising, sponsorship packages. In Italy in the last five years many families have decided to subscribe to Sky Sport to follow live soccer matches and every day the life of soccer players (there are three dedicated channel resp. for Milan, Inter and Rome).

Individual investors are more interested in investing in the stock market in football stocks because interest in the game has increased and they argue that there is a relation between sporting results and stock market performance. In this way the investors follow the sporting events as supporters and as stockholders, but this is not always the case, e.g., In Italy, when Rome team won the Italian championship, the supporters were crazy for the sporting event, but the stock (listed on Italian Stock Exchange went down because the company had to pay prize in money to the players.[9]

Some authors[10] have found empirical evidence of a link between international football results and the stock markets returns. They show a strong negative stock market reaction after football losses. Obviously this effect is higher in countries where football is economically significant. They argue that sports and soccer in particular are the most popular weekend activities, both in terms of active participation and passive observation, and it could have an important effect on Monday's opening stock prices.

It is true that also other events linked to football companies could weigh on the stock value. In Italy from May to July 2006 the Football scandal burst and two teams listed in Italian Stock Exchange were inquired (Juventus and Lazio). It was clear that even if Juventus won the Italian Championship in May, the stock in market went down because the first and the second sentence condemned them to Serie B (as First Division in England). The economic damage is high because the company will lose the availability to stipulate new contracts and to waive a revenue (many millions in euros) from European Champion League. The sporting

---

[8] Deloitte, 2004, "Annual review of Football finale", Sports Business group at Deloitte, Manchester, England)

[9] Edmans, A., Garcia D., Norli, O., 2005, "Football and stock returns", Working paper, www.ssrn.com

[10] Deloitte, 2004, "Annual review of Football finale", Sports Business group at Deloitte, Manchester, England

damage is equally high because it is the first time in its history that Juventus will play in Serie B and many famous players preferred to play in other famous European teams (Barcelona, Real Madrid, Inter). After these sentences many investors argue that it's not a good investment to buy football stocks because they are too closely linked to sporting results and too much influenced by the mood of investors.

Other studies show that the consumption choices are determined by sporting results, in fact if the results are negative there will be a negative reaction on consumer demand. Falter, Perignon and Vercruysse[11] have considered experiments based on one of the most important sporting events in the world: the Soccer World Cup (French 1998). They tested the impact of a World Cup Victory on the demand for Soccer in this country, they have identified the period following a Soccer World Cup victory as a period of overwhelming joy for the winning country. In literature it is shown that consumer sentiment is a key determinant of the demand for any good or service; they argue that there is a clear influence of positive mood on consumer behaviour using soccer data.

There is strong evidence in the literature[12] which shows the importance of soccer for the economies of some countries.

It will be interesting to observe if in Italy there will be a good impact on the consumption demand after 9 July 2006 (the day of the Cup final) or the football scandal after the second sentences will contribute to reduce the interest in football affairs.

Also in a Moslem country like Turkey soccer is very important in the mood of individuals, an interesting study Berument Ceylan Gozpinar assessed the effect of soccer success on stock market returns. The empirical evidence suggests that Besiktas's victory against foreign rivals in the Winner Cup increased the stock market returns. The same effect is not present for the other two big teams (Fenerbahce and Galatasaray).

At the end of this first part we may wonder how convenient is – during the stock picking – to choose football stocks for our portfolio. Many research papers show some kind of relationship between sport performance and financial performance, therefore you should account for some volatility in returns driven by the green field results and we could also maintain that, as for other industry sectors, you need proper football expertise to invest in a listed soccer team.

3 EMPIRICAL ANALYSIS: METHODOLOGY

The objective of this analysis is to assess the effect of soccer success on stock market returns. Particularly we are interested to assess this effect when the soccer clubs themselves are publicly listed.

In this case, the "emotional effect" on stock market may be grounded on a somewhat rational basis. By showing good performance, a club may raise more funds, mostly through sponsorship and media contracts, but also by selling tickets or merchandise to supporters.

When sport performance impacts prices of a listed soccer team, we can distinguish two transmission channels.

First we have the emotional channel. Investor sentiment plays a key role in determining their investment decisions. According to Stracca it is "plausible that some emotional factors play

---

[11] Falter, J., M., Perignon C., Vercruysse O., 2005, "Impact of overwhelming joy on consumer demand: the case of a soccer World-Cup victory" Working Paper

[12] Dobson, S., Goddard, J., 1998, "Performance, revenue and cross subsiditiation in the football league 1927-1994", Economic History Review 4 pp. 763-785.
Pollard, P. S., 2002, "Groowwwth!", International Economic Trends Annual Edition, July.
Berument, H., A. Inamlik, Yucel, E. M., 2003, "The effect of soccer on productivity (in Turkish)", Iktisat, Isletme ve Finans, 212, pp 51-62.
Berument, H., Yucel, E. M., 2005, "Long Live Fenerbanche: Production Boosting Effects of Soccer in Turkey", Journal of Economic Psycology.
Berument, H., Ceylan, N. B., Gopinar, E., 2005, "Performance of Soccer on the stock market evidence from the Turkey", Working Paper.

an important role in the setting of financial market prices. Especially the emotional states associated to 'feeling good' or 'feeling bad' may be important to the extent that they influence expectations formation."[13]

In this case we use the expression "emotional effect" to mean the fact that, despite investors' decisions may result in a positive economic result, their main motivation is not strictly an economic one or at least the decisions are not based on the sole financial calculus. We could quote Keynes 's "animal spirit" to justify the human heuristic ability to solve complex investment challenges.[14]

A second impact on financial performance is driven by potential or actual profits due to positive sport performance. This may be regarded as a more standard effect: consumer tend to buy goods and services that they like more based on product ability to satisfy their economic needs. Speaking of soccer teams, the quality of their products may be measured by their game performance. To put it in other words, we may say that, since soccer plays are not theatre plays, the consumer's need as a fan can be not just to watch the match, but to see his or her team winning the match.

Indeed, when a company is a sponsor of a team it is because it expects to gain notoriety and to sell more; it is rather natural to expect similar effects apply to the products of the sponsored team itself: as a general rule, for example, the more a team is notorious and winning the more a TV network will pay to buy broadcating rights.

If prices and volumes for products sold by a winning team are higher, that will result in higher profits for their investors: this is a key driver in portfolio selection decisions.

There may be many indicators of soccer performance: attendance of international competition, number of cups gained by the clubs, ranks provided by national soccer leagues etc., but of course the most obvious one is given by the goal scored during matches played. This is also a very convenient one because it is publicly available and is updated frequently. In fact an important club can play even twice a week in domestic or international competitions, this is important to us in comparing sport performance with financial performance, because we can immediately check, for example, if the emotional effect of a victory has an impact on related stock quotes.

One draw back of this indicator is that it could be a very volatile one. After all, soccer is a game (of chance too) and therefore there may be a random component in the final score of a single match. It could actually happen that a team ranking last in a reference chart may win against a team ranking first.

Indeed there could be performance indicators more effective in assessing stable, long run performance of a soccer club, but, if they are not linked to attention grabbing events, they may fail to capture and affect the mood of the emotional investor, and, if they are too infrequent, they have scant significance in a formal analysis[15].

After all, the chance and the unforeseen, that accompany a match, can play an important role in attenders' emotional process, and this may be well captured by the final score.

Based on this observations we turned to use scores as the target variable for clubs' sport performance, in three different modality:

- *net scores*: that is, the differences between goals scored and goal against;

---

[13] Stracca, L., 2003, "Behavioral finance and asset prices: Where do we stand?", Journal of Economic Psychology, 1-33.

[14] "Heuristics are rules of thumb, or mental shortcuts, the human brain uses to quickly solve complex problems. For example, a billiard player does not solve the trigonometric and differential equations needed to determine at what angle and speed to hit the cue ball in order to put another ball in the correct pocket. Rather, a billiard player uses rules of thumb and mental shortcuts that allow him to play the game, even though he may not understand the mathematics.", Fuller, Russell J., 2006, "Behavioral Finance and the Sources of Alpha", RJF Asset Management.

[15] For example winning of important international soccer competition may have a very strong impact on supporters' mood as investors, but these are also by definitions rare events for all clubs and as such not suitable to base our analysis on.

- *wins*, *ties* and *losses*: which may be regarded as the categorical version of the net score variable;
- *wins*, *non-wins*: the dichotomous version of the previous modality;

Coming to the other side of the performance, the financial performance, we perhaps move on a somewhat safer ground as there is a wide literature on the subject.

An important aspect, that need to be stressed, is that the soccer case may be regarded as an application of the event study theories. In this case the piece of news is given by the match result, possibly the success for the listed club analysed. A relevant point here is that, as a general rule, news is considered "as is". In a scientific study it is not easy or feasible to assess the quality of the information: i. e. distinguish between good news and bad news. But soccer — or sport in general — is a special case, because a success is easy understood as a good news event and conversely an unsuccessful game[16].

Based on our previous considerations, we are interested in assessing the impact of soccer scores of a listed club on its quotes. If there is a positive impact, we expect that on the day following winning matches prices will go up and vice versa after lost matches.

To these ends, post match prices/returns are compared with their average on the previous five trading days.

We do not just compare these prices with one day lagged price: this is first intended to smooth the noise that we may have on one single day; secondly, given their leisure nature, many matches happen on weekends, therefore the next trading days happen typically on Monday, it is interesting in this case to compare the price of the starting trading week with average price in the preceding week, in consideration of the match event in the middle.

We now formalise these concepts.

Assume that s_{im} is the net score the i-th club gets playing on the date τ_m and p_{ij} is the price of its listed stocks on the on the date t_j.[17]

We give the usual definition for returns, as the logarithmic first difference of the prices:

$$r_{ij} = (\ln(p_{ij}) - \ln(p_{i,j-1})) * 100$$

We use the symbol $\diamond$ as the *next trading day operator*, such that if τ_m is a match date, $t_{\diamond m}$ is the date of next trading day following τ_m. It follows that $p_{i\diamond m}$ is the price on trading day $t_{\diamond m}$, following τ_m — relative to i-th club.

Now, given the match played in τ_m by the i-th club, we define the *Post Match Delta Return* (PMDR) as:

$$\delta_{im} := r_{i\diamond m} - \frac{1}{a}\sum_{h=-a}^{-1} r_{i,\diamond m+h}$$

So, with reference to i-th club, δ_{im} expresses the i-th return variation observed on business day following τ_m, with respect to the average return observed in the a (trading) days before.

To better understand the motivation underlying this modelling, we present an example based on Table 1.

Suppose we have three fictional stocks, A, B and C, each referred to soccer clubs, which have the same price on trading day 6 (Friday) and 7 (Monday), and suppose that all the three clubs won a match during the weekend in the middle. Given the steep price increase on Monday, we attribute this to the sport performance.

There are but some differences. A showed already a previous positive trend, B on the contrary reverted with day 7 a previous negative trend, finally C showed a really flat performance in days before, always the same price.

---

[16] On this point see Boido, C., Fasano, A., (2006). "News ,Prices, Volumes"

[17] Note that matches and prices are indexed to different dates. This means that, if you take e. g. the fifth element of each series (s_{i5} and p_{i5}), they are likely to refer to different dates and this is also why we use distinct notation for them (resp. τ_5 and t_5).

Table 1. Average Return Before matches (ARBM) and Post Match Delta Return (PMDR)

Trad. Days	A Price	A Log-Ret.	B Price	B Log-Ret.	C Price	C Log-Ret.
1	0.9		1.9		1.4	
2	1	10.54	1.80	-5.41	1.40	0.00
3	1.1	9.53	1.70	-5.72	1.40	0.00
4	1.2	8.70	1.60	-6.06	1.40	0.00
5	1.3	8.00	1.50	-6.45	1.40	0.00
6	1.4	7.41	1.40	-6.90	1.40	0.00
7	1.7	19.42	1.70	19.42	1.70	19.42
ARBM		8.84		-6.11		0.00
PMDR		10.58		25.52		19.42

Previous trend is important to us, because a price increase after an overwhelming victory is more significant if it was able to invert a previous trend.

Now, as you can check by the figures in table, log return in day 7, is always the same (19.42). By the way, by subtracting the average return before matches (ARBM the in table): we mitigate the effect in for A, to reflect the fact that a bull trend was already operating; on the contrary we amplify the effect for B, so stressing the importance of trend inversion; finally, in the case of C stock, the stable previous prices involve neutrality and the PMDR is the same of the the log-return.

Once defined these quantities, the objective is to study, for the given clubs, the statistical relationship between δ_* and s_*.

On the basis of our behavioural tenets, we expect the higher the net scores the higher the post match variations, that is $\Delta s_{im} > 0 \Rightarrow \Delta \delta_{im} > 0$.

As observed, we can categorise s_* in terms of 'wins', 'tie' and 'loss' or simply via the dichotomous 'win' and 'non-win'.

It is particularly useful for statistical treatment to the define dichotomous variable:

$$w_{im} =: \begin{cases} 1 & \text{if } s_{im} > 0 \\ 0 & \text{otherwise} \end{cases}$$

Through w_{im} we can split PMDRs in the two groups:

$$D_i^+ := \{\delta_{im} : w_{im} = 1\},$$
$$D_i^- := \{\delta_{im} : w_{im} = 0\},$$

which express the variations following won matches only (D_i^+) or conversely variations for matches lost or ended in tie (D_i^-).

Clearly, if the impact of scores on listed prices is true, we expect a difference in mean between the two sets.

Dataset: Coming to the data set, it concerns the publicly listed soccer clubs in Italy, namely Juventus, Lazio and Roma. The data is constituted of closing prices and scores for the period ranging from January 2005 to June 2006.

Financial data is from Bloomberg provider, while scores were provided by Global Sports Media.

The matches played during the reference period were:

Club	Matches played
Juventus	52
Lazio	53
Roma	52

These are to be referred to 345 observations relative to clubs' stock prices.

Note that, despite soccer data was available categorised for different type of sport competition (domestic and international), to avoid subjective bias, we did not weighted the importance of single matches in any way.

The data was treated with a custom application developed in R[18]. Commented source code is available in the Appendix to this paper. Comments embedded in the code contain many implementations detail that may be of interest to the reader.

The results are generated both in a tabular and graphical format and are discussed in next section.

4 EMPIRICAL ANALYSIS: OUTPUT

We start by giving summary statistics relative to scores and PMDR.

```
Summary data for   Juventus
       PMDR                 Scores
Min.    :-14.2116   Lost  : 3
1st Qu.:  -0.6899   Drawn:12
Median :   0.2018   Won   :37
Mean   :   0.2096
3rd Qu.:   1.6073
Max.   :   7.7466

Summary data for   Lazio
       PMDR                 Scores
Min.    :-18.32710  Lost  :15
1st Qu.: -1.71083   Drawn:16
Median :  0.04994   Won   :22
Mean    : -0.49242
3rd Qu.:  1.41846
Max.    :  7.84742

Summary data for   Roma
       PMDR                 Scores
Min.    :-5.4900    Lost  :16
1st Qu.:-1.6607     Drawn:18
Median :-0.3322     Won   :18
Mean    :-0.5310
3rd Qu.: 0.5078
Max.    : 3.2329
```

We see the narrowest range in return variations for Rome, while the highest net scores (35) is on behalf of Juventus.

PMDRs can be compared visually in figure 1 through the use of boxplots[19].

We see that Juventus financial performance after the playing has a good symmetry, while Lazio and Rome distributions are slightly asymmetric.

To assess visually the relationship between financial and sport performance we can check figures 2, 3, 4,

As a rule we should expect quadrants A and D to be less crowded than B and C.

One question arises as to the impact of draw. Here data seems to show that markets dislike matches that end in ties. This means that the dichotomous 'win' or 'non-win' variable is a better choice to model performance.

We check this observations numerically too, through contingency tables.

Return vs. Performance

---

[18] R is an open-source language and environment for statistical computing. See http://www.r-project.org for details.

[19] A boxplot is a way to look at the overall shape of a set of data (center, spread, skewness, and outliers). The central box extends between the quartiles so that it contains the middle half of the data. The line in the middle of the box represents the median, and by its positions one can assess the distribution symmetry.
"Whiskers" generally extend out of the box for 1.5 times the inter-quartile range, and they let identify easily the outliers which locate themselves out of the "fences".
For more see Chambers, J. M., Cleveland, W. S., Kleiner, B., Tukey, P.A., (1983), Graphical Methods for Data Analysis. Wadsworth & Brooks/Cole.

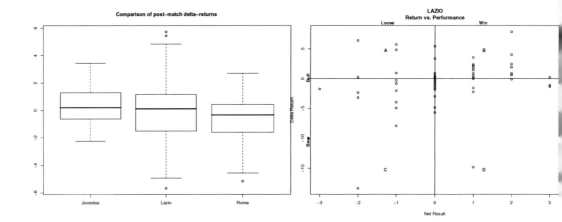

Fig. 1. Boxplot of post-match variation of returns

Fig. 3. Sport Performance vs. Financial Performance: Lazio

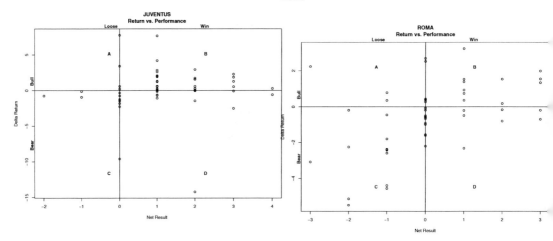

Fig. 2. Sport Performance vs. Financial Performance: Juventus

Fig. 4. Sport Performance vs. Financial Performance: Roma

```
Contingency table for:   Juventus

                   Results
PMDR Classes     Non-Win  Win
[-18.3,-1.75]          5    8
(-1.75,0.0236]         3   10
(0.0236,1.6]           4    9
(1.6,7.85]             3   10
       Total:         15   37
```

```
Return vs. Performance
Contingency table for:   Lazio

                           Results
PMDR Classes           Non-Win  Win
[-14.2,-0.671]               9    5
(-0.671,0.204]               6    7
```

(0.204,1.57]	6	7
(1.57,7.75]	10	3
Total:	31	22

Return vs. Performance
Contingency table for: Roma

	Results	
PMDR Classes	Non-Win	Win
[-5.49,-1.71]	12	1
(-1.71,-0.332]	10	3
(-0.332,0.584]	8	5
(0.584,3.23]	4	9
Total:	34	18

Returns are grouped on the basis of the Tukey five numbers (used also in the summary statistics).

There is a few data inconsistent with the behavioural hypothesis, which represent points that populate quadrants A and D, but generally incoherent observations appear to be less numerous then coherent ones.

Some figures such those of a winning Juventus on playing fields and losing on financial markets, seem a bit anomalous.

To check for Significance we now turn to Student t-tests. To this end we split PMDRs in two groups, based on success dual variable, as shown in the previous formalisation (D_i^+ amd D_i^-).

t-test for Juventus

```
        Shapiro-Wilk normality test

data:  Win
W = 0.6818, p-value = 1.121e-07


        Shapiro-Wilk normality test

data:  Non-Win
W = 0.8179, p-value = 0.006298


        Welch Two Sample t-test

data:  Win and Non-Win
t = 1.0618, df = 22.893, p-value = 0.8503
alternative hypothesis:
  true difference in means is less than 0
95 percent confidence interval:
     -Inf 2.891518
sample estimates:
  mean of x   mean of y
  0.5286470 -0.5773308
```

t-test for Lazio

```
        Shapiro-Wilk normality test

data:  Win
W = 0.7331, p-value = 5.314e-05


        Shapiro-Wilk normality test

data:  Non-Win
W = 0.8652, p-value = 0.001087


        Welch Two Sample t-test

data:  Win and Non-Win
t = 1.325, df = 48.604, p-value = 0.9043
alternative hypothesis:
  true difference in means is less than 0
95 percent confidence interval:
     -Inf 3.578946
sample estimates:
  mean of x   mean of y
  0.4316011 -1.1481696
```

t-test for Roma

```
        Shapiro-Wilk normality test

data:  Win
W = 0.9728, p-value = 0.8483


        Shapiro-Wilk normality test

data:  Non-Win
W = 0.9619, p-value = 0.2762
```

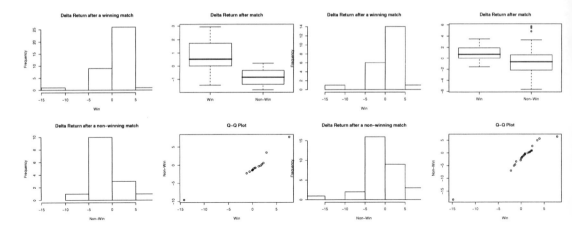

Fig. 5. Comparison of win/non-win financial performance: Juventus

Fig. 6. Comparison of win/non-win financial performance: Lazio

```
            Welch Two Sample t-test

data:  Win and Non-Win
t = 3.6125, df = 47.803, p-value = 0.9996
alternative hypothesis:
  true difference in means is less than 0
95 percent confidence interval:
     -Inf 2.415855
sample estimates:
 mean of x   mean of y
 0.5477222  -1.1020955
```

The average performance, in terms of return variation after winning matches, appears in all the three cases positive, while, after unsuccessful matches, the average is negative. This is consistent with theory, but, as for the significance, p-value appears too high. By the way these results may be in the case of Juventus and Lazio be the effect of non normality.

The comparison of win/non-win is also presented in a visual form in figures: 5, 6, 7.

The plots confirm the asymmetry of the sample distributions, particularly boxplot — higher on the winning side — imply a positive impact of sport performances on financial ones.

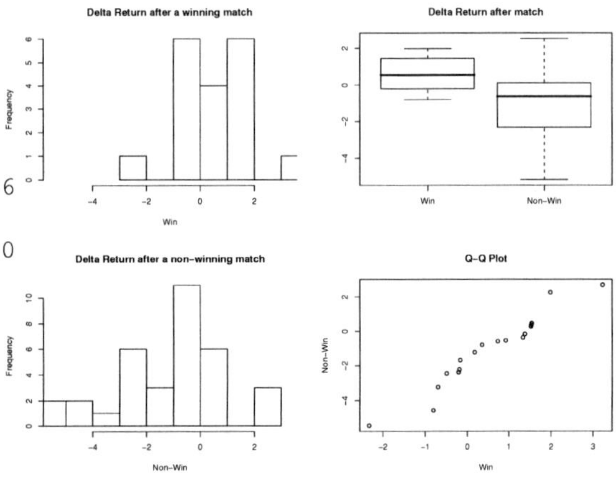

Fig. 7. Comparison of win/non-win financial performance: Roma

To cope with statistical significance and given the samples' asymmetry, we resorted to non-parametric techniques.

To this end for each team, we extracted randomly, from all the possible PMDRs relative to our reference period (i.e. 345 observations), a number of variations that equalled the matches played. The idea behind is that a generic sample mean should be comprised between the lower bound of the mean of the "never-won-matches" sample and the upper bound of the "won-matches-only" sample.

```
Sampling:  Juventus
    Sample size:  52  out of  345
    Trial     Mean              >NoWin mean    <Win mean
     1        0.02405845        TRUE           TRUE
     2        0.05655788        TRUE           TRUE
     3       -0.07310237        TRUE           TRUE
     4       -0.01009738        TRUE           TRUE
     5        0.08288204        TRUE           TRUE
     6        0.06743323        TRUE           TRUE
     7        0.04378302        TRUE           TRUE
     8        0.06956866        TRUE           TRUE
     9       -0.02014645        TRUE           TRUE
    10        0.07387201        TRUE           TRUE
All samples mean:  0.03148091
Post Win mean:   0.528647
Post Non-Win mean:  -0.5773308

Sampling:  Lazio
    Sample size:  53  out of  345
    Trial     Mean              >NoWin mean    <Win mean
     1        0.1096848         TRUE           TRUE
     2       -0.007291506       TRUE           TRUE
     3       -0.07102549        TRUE           TRUE
     4       -0.08257605        TRUE           TRUE
     5        0.07766938        TRUE           TRUE
     6       -0.03203867        TRUE           TRUE
     7        0.06543365        TRUE           TRUE
     8       -0.1102873         TRUE           TRUE
     9       -0.03200535        TRUE           TRUE
    10        0.02105164        TRUE           TRUE
All samples mean:  -0.006138485
Post Win mean:   0.4316011
Post Non-Win mean:  -1.148170

Sampling:  Roma
    Sample size:  52  out of  345
    Trial     Mean              >NoWin mean    <Win mean
     1       -0.005721752       TRUE           TRUE
     2        0.03343388        TRUE           TRUE
     3       -0.0647569         TRUE           TRUE
     4       -0.05735351        TRUE           TRUE
     5       -0.02963383        TRUE           TRUE
     6       -0.02567209        TRUE           TRUE
     7       -0.04160218        TRUE           TRUE
     8       -0.02314614        TRUE           TRUE
     9        0.0002006721      TRUE           TRUE
    10       -0.02654829        TRUE           TRUE
All samples mean:  -0.02408001
Post Win mean:   0.5477222
Post Non-Win mean:  -1.102096
```

As you can see, samples means extracted are always higher of the mean past unsuccessful matches and are lower than the mean past victories.

5 CONCLUSION

Soccer — and sport in general — can be one of the most relevant factor to influence investors' mood analysed by behavioural finance researchers. Successful sport competition may be also regarded as a special application for the literature on the event study, where the event is given by the winning performances. The peculiarity for sport is that, while in general a news release cant be weighted as good or bad news in an objective manner, this is quite trivial in the sport field, where it is easy to qualify a win or a good sport performance as a good news for the involved club regarded as a profit company.

Another relevant aspect is that in this case emotional and economic motivations can join together. Indeed, if the national team wins an important competition, this may have an optimistic effect on consumer demand and financial behaviour of investors, besides, when a listed club is a winner on the playing field, this involves not only the moods of its supporters and investors, but will also boost ticket sales as well as sponsorship and media contracts.

In our analysis we considered this aspect by studying the series of scores and market prices together, with respect to Italian publicly listed clubs— currently only three.

Our data shows, for all the involved clubs, that the average prices/returns immediately after wins are higher than average prices/returns after unsuccessful matches. It is important to stress that we did not examined some isolated cases connected to special events or competitions, but all available data on scores for matches played during a year and half. Particularly we examined also the impact of lost and draw matches, to check for possible inconsistent results.

It seems on this side that Italian investors dislike matches ending in ties.

As for statistical significance, parametric test does not show satisfactory results. also because of the non normality of data. But the fact itself that for all the teams we get coherent result consistent with our hypothesis is a sign stability.

However this problems are easily overcome by random sampling data. Prices/returns samples generated by random extractions have means neither higher than mean observed after winning matches nor lower than mean observed after lost and draw matches.

Based on this result we feel confident to assert that for Italian markets the sport performance affects the financial performance of listed soccer clubs.

6 REFERENCES

[1] Berument, H., A. Inamlik, Yucel, E. M., (2003). "The effect of soccer on productivity (in Turkish)", Iktisat, Isletme ve Finans, 212, pp 51-62.

[2] Berument, H., Yucel, E. M., (2005). "Long Live Fenerbanche: Production Boosting Effects of Soccer in Turkey", Journal of Economic Psycology.

[3] Berument, H., Ceylan, N. B., Gopinar, E., (2005). "Performance of Soccer on the stock market evidence from the Turkey", Working Paper.

[4] Boido, C., Fasano, A., (2005). "Calendar Anomalies: Daylights Savings Effect" The ICFAI Journal of Behavioural Finance vol II n. 4 december pag 7-24.

[5] Boido, C., Fasano, A., (2006). "News ,Prices, Volumes: The Behaviour of small cap investors on Italian Stock Exchange", 30th Anniversary of The Journal of Banking and Finance Conference Peking University Beijing China 6-8 June . Working paper.

[6] Boido, C., Fasano, A., (2006). "Small Caps vs. Big Caps: a comparison based on the impact of market news", 26 th International Symposium on Forecasting Santander, 11-1.4 June Working paper.

[7] Buehler, R., Grifin, D., Ross, M., (1994). "Exploring the planning fallacy: Why people underestimate their task completion times", Journal of Personality and Social Psychology, 67, 366-381.

[8] Carroll, D., Ebrahim, S., Tilling, K., Macleod, J., Smith, G., (2002). "Admission fro myocardial infarction and World Cup football:database survey", British medical Journal, 325 p 1439-1442.

[9] Chambers, J. M., Cleveland, W. S., Kleiner, B., Tukey, P. A., (1983). Graphical Methods for Data Analysis. Wadsworth & Brooks/Cole.

[10] Deloitte (2004). "Annual review of Football finale", Sports Business group at Deloitte, Manchester, England.

[11] Dobson, S., Goddard, J., (1998). "Performance, revenue and cross subsiditiation in the football league 1927-1994", Economic History Review 4 pp. 763-785.

[12] Edmans, A,. Garcia, D., Norli, O., (2005). "Football and stock returns", working Paper, www.ssrn.com.

[13] Edward, W., (1968). "Conservatism in human information processing", in B. Kleinmutz ed., Formal Representation of Human Judgement, New York, John Wiley and Sons.

[14] Falter.J.,M., Perignon, C., and Vercruysse O. (2005). "Impact of overwhelming joy on consumer demand: the case of a soccer world-Cup Victory", working paper.

[15] Fischhoff, B., Slovic, P., Lichtenstein S., (1977). "Knowing with certainty: the appropriateness of extreme confidence", Journal of experimental Psicology: Human perception and performance, 3, 552-564.

[16] Fuller, Russell J., 2006, "Behavioral Finance and the Sources of Alpha", RJF Asset Management.

[17] Gervais, S., Odean, T., (2001). "Learning to be overconfident", Review of Financial Studies, 14, 1-27.

[18] Hirshleifer, D., Shumway, T., (2003). "Good day sunshine :stock returns and the

weather", Journal of Finance, 58 (3) p 1009-1032 show that even if the weather change, the average daily return increases by only 9 basis points.

[19] Kahneman, D., Tverski, A., (1974). "Judgement Under uncertainty: heuristics and biases", Science, 185, 1124-1131.

[20] Kahneman, D., Tversky, A., (1979). "Prospect theory: an analysis of decision under risk" Econometrica 46, 171-185.

[21] Markman, K.D., Hirt, E.R., (2002). "Social prediction and the allegiance bias", Social Cognitive, n.20 pp. 58-86.

[22] Pollard, P. S., 2002, "Groowwwth!", International Economic Trends, Annual Edition, July.

[23] Stracca, L., 2003, "Behavioral finance and asset prices: Where do we stand?", Journal of Economic Psychology, 1-33.

[24] Wann, D., Melcnick, M., Russel, G., Pease D. (2001). "Sport fans: the psychology and social impact of spectators, Routledge.

[25] Weinstein, N., (1980). "Unrealistic Optimism about Future Life Events", Journal Of personality and Social Psychology, 39, 806-820.

Workshop:

Technical Applications in Medicine

Chairman: Martin Zauner

Scientific Board:
Karl Bögl, Actelion Pharmaceuticals Ltd.
Heinz Brock, AKH Linz
Reinhard Hainisch, FH OÖ Campus Linz
Thomas Haslwanter, FH OÖ Campus Linz
Andreas Lindbaum, FH OÖ Campus Linz
Bernhard Quatember, Medizinische Universität Innsbruck
Kurt Schilcher, FH OÖ Campus Linz
Harald Schöffl, BioMed
Andreas Schrempf, FH OÖ Campus Linz
Martin Zauner, FH OÖ Campus Linz

A New Method to Determine the Sustainability of the Increased Watersoluble Antioxidative Status (ACW) in Tear Fluid Obtained without Stimulation after Iodide-treatments in Bad Hall

Sirid Griebenow[1], Gebhard Rieger[1], Jutta Horwath-Winter[2], Otto Schmut[2]

[1]Paracelsus-Gesellschaft für Balneologie und Jodforschung, Dr. Karl-Renner-Str.6, 4540 Bad Hall, AUSTRIA
[2]Universitäts-Augenklinik Graz, Medizinische Universität Graz, Auenbrugger Platz 4, 8036 Graz, AUSTRIA

ABSTRACT

Motivation: The tear fluid contains antioxidative protective mechanisms. By the attack of free radicals, arising by influence of ozone, UV light, smog, cigarette smoking etc., these antioxidative protective mechanisms can be destroyed. The so-called environmental induced dry eye can arise by the damage of the tear-fluid compounds by oxidative stress. Since mostly patients suffering from dry eye do not have much tear liquid, usually it is necessary to stimulate tear production for investigation. Now a new method is introduced: the measurement by means of photochemoluminescence (PHOTOCHEM, Analytik Jena AG, Deutschland). This method needs only 1-2 µl to determine antioxidative parameters, making it possible to investigate tear liquid without stimulation.
The sustainability of the increased antioxidative capacity was examined after iodide-treatments had been carried out.
Results: The ACW values in the tear fluid were still increased significantly 6 months after ophthalmo-iodine-iontophoresis therapy for patients with a three-week eye treatment duration.
Availability: The more than 6 months improvement of the antioxidative capacity in the tear liquid underlines the important value of the ophthalmo-iodine-iontophoresis treatment for patients with a dry eye condition.

Contact: paracelsus.gesellschaft@utanet.at

1 INTRODUCTION

Free radiacals can destroy tear film containing proteins, mucosubstances and lipids. A consequence is the so called „dry eye". Since mostly patients suffering from dry eye do not have much amounts of tear fluid for investigation, it is necessary to stimulate tear production for investigation.
Now a new method is introduced: the measurement by means of photochemoluminescence (PHOTOCHEM, Analytik Jena AG, Deutschland). This method needs only 1-2 µl to determine antioxidative parameters, making it possible to investigate tear fluid without stimulation.

2 METHOD

A world with ever constantly growing importance of radical developing environmental factors and increasing demands of health and food quality requires an accurate and fast method for determination of the antioxidative system. With the photochemoluminescence-method of the Photochem a very time- and cost-effective system is now beeing available. Photochem allows the determination of the integral antioxidative capacity of most diverse substances mixture, of individual antioxidants and of the superoxide dismutase. The application of the Photochem range from food-technological, chemical, pharmaceutical as well as agricultural analysis to biochemical and medical research and routine.

2.1 Chemiluminescence in measurement of antioxidant potential

Chemoluminescence is a potentially sensitive method. It allows for evaluation of end products of radical attack, observation of the reaction kinetics and determination of antioxidant potential based on radical scavencing capacity.

The method of chemoluminescence uses luminol (5-Amino-2,3-dihydro-1,4-phthalazine) which is excited and reacts with free radicals and poduces a photoemission. The oxidation of luminol is accompanied by striking emission of light at 424 nm.

Photochem is a device designed to measure antioxidant activity of pure compounds as well as extracts of biological materials. It uses chemoluminescence to measure free radical scavenging capacity. Photochem has been designed for the analysis of both lipid- and water soluble antioxidants.

A standardized volume of photosensitizer substance is added to the assay medium. The photosensitizer is optically excited to produce superoxide anion radicals. The sample (antioxidative compound) scavenges a part of superoxide anion radical. The remaining radicals are quantified by luminescence generation. Antioxidants are quantified by comparison with a standard. Trolox and ascorbic acid are used as standards.

This method offers a quick measurement (a single measurement takes less than 3 min). It is highly sensitive and requires nM concentrations.

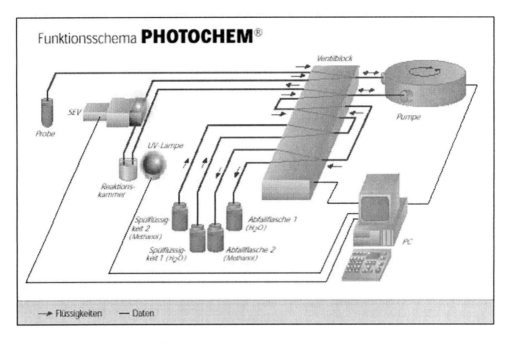

Fig. 1. Functional scheme of Photochem.

3 INVESTIGATIONS

A previous study has shown that the antioxidatve status can be positively influenced by the supply of the oxygen radical scavenger iodide taken up in the course of a cure in Bad Hall. The sustainability of the increased antioxidative capacity was examined after ophthalmo-iodine-iontophoresis-treatments had been carried out.

For the investigation of sustainability 21 patients after 6 months of iodine therapy were measured out of a group of 23 patients.

3.1 Results

The ACW values in the tear fluid were still increased significantly 6 months after iodine therapy for patients with a three-week eye treatment duration. The method of photochemoluminescence is a reliable method to determine the water soluble antioxidant capacity of little amounts of tear fluid and to observe the progress of the therapy.

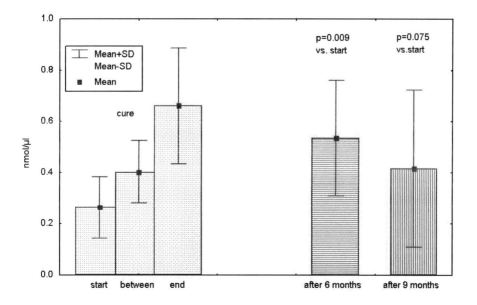

Fig. 2. Water-soluble antioxidant capacity of tear fluid.

4 REFERENCES

[1] Schmut, O.; Gruber, E.; El-Shabrawi, Y.; Faulborn, J. (1994) *Destruction of human tear proteins by ozone.* Free Rad Biology Med, 17/2, 165-169.

[2] Schmut, O. and Horwath-Winter, J. (2002) *The effect of sample treatment on separation profiles of tear fluid proteins: Qualitative and semiquantitative protein determination by an automated analysis system.* Graefe's Arch Clin Exp Ophthalm,l, 240, 900-905.

[3] Griebenow,S.; Rieger,G.; Horwath-Winter,J. (2003) *Eine neue Methode zur Bestimmung der antioxidativen Kapazität in der Tränenflüssigkeit.* Spektrum Augenheilk, 17 (5), 227-223.

[4] Kar Man Choy,C.; Cho,P.; Chung,W.Y.; Benzie,I.F.F. (2001) *Water soluble antioxidants*

in human tears: *Effect of the collection method.* Invest Ophthalmol Vis Sci, 42 (13), 3130-3134.

[5] Rieger,G.; Griebenow,S.; Winkler,R.; Stoiser,E. (2000) *Der antioxidative Status (TAS) vor und nach kombinierten Kurbehnadlungen in Bad Hall.* Spektrum Augenheilk, 14 (6),319-324.

[6] Horwath-Winter,J.; Schmut,O.; Haller-Schober,E.M.; Gruber,A.; Rieger,G. (2005) *Iodide iontophoresis as a treatment for dry eye syndrome.* Br J Ophthalmol, 89, 40-44.

[7] Griebenow,S.; Rieger,G.; Horwath-Winter,J.; Schmut,O. (2006) *Nachhaltigkeit der Erhöhung des wasserlöslichen antioxidativen Schutzmechanismus (ACW) in der nicht stimuliert gewonnenen Tränenflüssigkeit nach Iodid-Iontophoresebehandlungen in Bad Hall.* Spektrum Augenheilk, 20/1, 6-8.

CONCEPTS of Biomechatronics:

Individualized Prevention of Hearing Loss in the Working Environment and in Daily Life

Alexander Müller [a], Ulrike Fröber [a], Stefan Lutherdt [a], Hartmut Witte [a]

[a] Department of Biomechatronics, Faculty of Mechanical Engineering, Technische Universität Ilmenau, Max-Planck-Ring 12, 98692 Ilmenau, GERMANY

ABSTRACT

Motivation: Hearing loss is a big problem of civilization. Many of the sounds and noises surrounding people are too loud and endanger their auditory systems. It is a challenge for biomechatronics to develop mechatronical systems for the prevention of hearing loss.

Results: In future there should be available a miniaturized technical device for measuring stress and strain in the auditory system. This device should be able to protect persons in the working environment and in daily life.

Availability: The results of surveys and tests are available and published. The design and modeling phase started and it is the aim to have an exemplary realization at the end of our research.

Contact: alexander.mueller@tu-ilmenau.de
ulrike.froeber@tu-ilmenau.de

1 INTRODUCTION

In current ergonomics, for prevention of hearing loss, the usefulness of fixed and reproducible relations of psycho-physical stresses and resulting individual strains are generally underestimated. These relations often are hypothesized without being proved, and if they are proved their control in the working environment and in daily life is difficult.

At present, laws and ordinances for noise prevention exclusively define dB(A)-weighted limit values. In German federal pollution protection law (BImSchG), by the frequency assessment curve "A" as the only judgment criterion the assessment of sound pressure level is established. The International Standardisation Organization (ISO) agreed on this measurement procedure some fifty years ago to reach a simplification of judgment levels and to internationally fix a simple method. According to the hearing level threshold identified, low bass and very high frequencies are weighed less intensely than medium frequencies. "A" weighed levels as expression for the physical stress however cannot describe sufficiently the sensory strains on humans. The weights of "A" assessment only apply to narrowband noises and are derived from the "40 phone curve" (i.e. "isophones"). Consecutively it is assumed, that correct weighing only results if these prerequisites are given.

To support ergonomics in solving this problem, biomechatronics may provide adaptive miniaturized dosimeters. Dosimeters are used for evidence-based prevention of damage so that reproduction of dose-effects may be avoided. Noise measuring systems are supposed to transform readings of the same external stresses for different individuals into a measure of individual strains. Future research on strain should concentrate on the simultaneous identification of measurements for disposition and the assessment of stresses by loading. Using the current technological progress in micro-systems engineering, biomechatronics is on its way to develop a "personalized miniaturized mechatronic dosimeter (PMD)" for the prevention of noise-provoked damages in hearing.

2 CONCEPT OF PMD

Phase 1: Identification of the relevant psycho-physical basis, ideally up to the level of "evidence based medicine"
Phase 2: Derivation and definition of a catalogue of technical requirements
Phase 3: Implementation

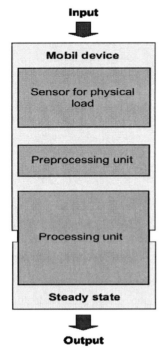

Fig. 1. Principle of the "Personalized Miniaturized Mechatronic Dosimeter" for arbitrary physio-chemical qualities and quantities.

The purpose of the project is to develop a procedure and an arrangement which allows for precise quantification of the individual stress and strain of an exposed employee. The focus is laid on a miniaturized device with personalized adaptation (e.g. CIC – completely in the canal) for each subject, and high data resolution (e.g. frontend) for the precise and comprehensive determination of binaural noise exposure or strain.

For the development of the PMD the results of other projects dealing with hearing can be used as a basis. During development of an assistance system for people with several handicaps, those with hearing impairments have been interviewed concerning their use of hearing aids, computers, and personal digital assistants (PDA). The following studies have been performed:

a) One-on-one interviews with acousticians, otorhinolarynglogists and concerned persons

b) Interrogations of people wearing cochelar implants or hearing aids and deaf people using the Delphi-method

c) Questionnaire campaigns

One of the main facts the surveys showed is that people with hearing loss that use a hearing aid often also use other technical devices. They do not have problems to use new devices and to learn how to use them.

A comprehensive database containing requirements, experiences and wishes of people with different kinds of hearing loss has been generated. From these results tasks and design criteria for the PMD also can be derived.

Fig. 2. Hearing impaired person while testing new technical devices.

Another important aspect of the PMD is to realize multifunctionality. Research in the field of noise recognition opens new opportunities to integrate noise recognition in the PMD. So it could be possible to use the dosimeter also for warning and alarm indication. It could work as an active protection for both healthy and diseased auditory systems.

In the future the PMD should offer a chance for early identification of people with over sensitive hearing ("vulnerable hearing").

3 ACKNOWLEDGEMENTS

We like to thank the „Berufsgenossenschaft Nahrungsmittel und Gaststätten" for their technical support while doing the experiments and for financial support.
Also we want to thank "Deutscher Schwerhörigenbund e.V." for excellent cooperation.

4 REFERENCES

[1] Fröber, U. ; Lutherdt, S. ; Koch, M. ; Witte, H. ; Kurtz, P. ; Roß, F. ; Wernstedt, J.: *TAS - Ein Touristisches Assistenzsystem für den barrierefreien Zugang zu Urlaubs-, Freizeit- und Bildungsaktivitäten im Thüringer Wald.* In: Proceedings of the 50th Internationales Wissenschaftliches Kolloquium, 19.-23.09.2005. ISBN 3-932633-98-9.

[2] Lutherdt, S. ; Fröber, U. ; Wernstedt, J. ; Witte, H. ; Kurtz, P.: *Development of assistance systems for user groups with specific handicaps – a challenge for the ergonomic design process.* In: Proceedings of the XIX Annual International Occupational Ergonomics & Safety Conference, 27.-29. Juni 2005. ISBN 0-9652558-2-4.

[3] *Exploration von Bedarfen und Leistungsanforderungen an ein Assistenzsystem aus Sicht von Menschen mit Hörbehinderung*, Aproxima (2006).

[4] Lipsius, P ; Emmerich, E. ; Grosch, J. ; Rudel, L. ; Walter, M.: *Musiker. Die ideale Untersuchungsgruppe für Lärmschäden.* In: Grieshaber, R.; Stadler, M.; Scholle, H.-Ch.: Prävention von arbeitsbedingten Gesundheitsgefahren und Erkrankungen, 11. Erfurter Tage. Jena: Verlag Dr. Bussert und Stadeler, 2005. – ISBN 3-932906-64-0, S. 505-515.

[5] Müller, A. ; Albrecht, B. ; Grosch, J. ; Stubenrauch, M.; Emmerich, E.; Mollenhauer, O. ; Schade, H.-P. ; Witte, H.: *Miniaturized dosimeters of an individualized prevention of noise-realted impairment in the working enviroment.* In: Biomedizinische Technik 49 (2004), Nr. 2, S. 1028-1029. – ISSN 0939-4990 .

Application of a Miniature Endoscope to Investigate the Inner Surface of Blood Vessels

Reinhard Hainisch[a]*, Stefan Froschauer[b], Harald Schöffl[b], Martin Zauner[a]

[a]Upper Austria University of Applied Sciences, Department of Medical Technology,
Garnisonstraße 21, 4020 Linz, AUSTRIA
[b]BioMed – Centre for biomedical and medical technology research,
Garnisonstraße 21, 4020 Linz, AUSTRIA

ABSTRACT

Motivation: An important factor for the outcome of microsurgical procedures is the quality of the micro vascular anastomosis. During the training of microsurgical procedures the inside of the resulting suture is commonly assessed through the dissection of the vessel. The inside-investigation of the anastomosis by the aid of a miniature endoscope was not feasible due to unsatisfying image quality.

Results: We have developed a method and an image-processing algorithm that allows the inspection of small diameter vessels with miniature endoscopes. With this kind of investigation, it is possible to evaluate the lumen after microsurgical anastomosis without the need of the dissection of this vessel.

The resulting single picture offers a new efficient kind of documentation for the investigation of the anastomoses. It shows in a 2-dimensional image, a map of the inner surface, which allows a quick and easy evaluation of the intervention.

Availability: The system was set up and tested with several combinations of flexible and rigid endoscope and camera setups. The image processing was executed in real-time on the development platform. Up to know we did not implement a user friendly runtime version that can be used by a surgeon.

Contact: reinhard.hainisch@fh-linz.at

1 INTRODUCTION

Microsurgical anastomoses often have diameters of less than one millimetre. The assessment of the patency of such a vascular suture adheres to well-defined criteria from a clinical point of view. It is important for the evaluation of an anastomosis to investigate it from the outside as well as from the inside of the vessel. During microsurgical training, there is a lack of objective control instruments to evaluate the sutured vessel without dissecting the vessel. However, the quality of such microsurgical anastomosis is important in complex surgical reconstructions. The possibility of an evalutation contributes to a better quality in treatment. A new endoluminal technique employing miniature endoscopes that can be inserted into the vessel may solve this problem. Figure 1 shows a typical image of the examination of a sutured blood vessel through a miniature endoscope. The suture is hard to see.

The aim of the project was a significant enhancement of the image quality from the flexible mini-endoscope in general. A particular focus was on the improvement of the evaluation quality of anastomosis after microsurgical interventions.

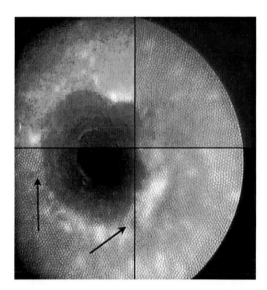

Fig. 1. Mini-endoscopic image of an anastomosis, which illustrates the honeycomb effect. The four quadrants show examples of possible image enhancement trough digital image processing. The arrows point at parts of the suture, which are of interest to the surgeon.

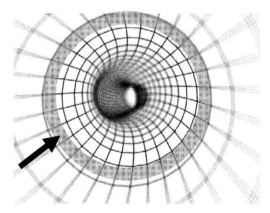

Fig. 2. Problem of low optical depth due to common lens systems. The arrow points on into an area marked by a grey circle where the image is sharpest, with proper illumination. The surgeon inspects usually this area because only there he can get the relevant information for evaluating the anastomosis. This is the area with the highest density of relevant information.

Physical limits in the construction of flexible mini-endoscopes mainly restrict in the image quality is. The image in figure 1 is dominated by a honeycomb effect. This is caused by the limitations in the production of the glass fibre bundle. An optical system of 1.0mm in diameter consists of approximately 10000 single fibres that have each a diameter of about 100 nm. Interstices arise between the fibres due to the round profile, the coatings and the order of the fibres. These areas cannot guide light to the camera. The size of the image of about 1mm² has to be compared to the amount of aligned glass fibres of about 100 in each direction. The ratio of dimensions of the investigated structures and its imaging modality result in a rough resolution that is finally seen in a honeycomb-effect.

Fig. 2 shows the problem of the very narrow optical depth of field caused by the optical properties of common lens systems. With the application of gradient index (GRIN) lenses, the depth of field can be improved. These optical elements have a gradient of the index of refraction throughout the volume of the lens, which allows the correction of the optical distortion.

Modern endoscope cameras deliver the images in digital format via Firewire-Port (IEEE1394) or USB. Therefore, it is possible to acquire the image directly digital from the CMOS-camera without signal distortion. Digital processing filters can further enhance the quality of the picture. Figure 1 shows an image that is partially processed with the optimisation of the histogram and the application of Gaussian filters for edge detection.

However, through improvements in the optical system and digital image processing we still get images of limited clinical relevance. It is necessary to obtain an image that permits a qualitative evaluation of a suture from the inside of a vessel.

A new approach analysing the viewed structure and the imaging modality was developed. A motion simulation model of the endoscope and its image during the inspection of a vessel was used to implement and test the image-processing algorithm.

2 METHODS

Important for the evaluation of the anastomosis is the perception of the investigated region. The quality of a single image is not always the key to an appropriate cognition of the area. It is the amount of information that an image delivers which makes it relevant. In our case, we investigate a hollow tube, like a blood vessel, in axial direction. In the middle of the image, the degree of information is relatively small because the illumination does not extend that far, and small details are hard to see in the distance. The area near the optical system on the margin of the image is mostly diffuse, dim, overexposed or too bright and therefore cannot be used for the evaluation of the suture. The area in a single image that contains the most relevant information lies along a circle. This circle can be recognized in light grey in Figure 2. Here it covers a part of the inner surface in the region where the optical system delivers the sharpest image.

New Approach: The round-scan system

An endoscope camera delivers 25 images per second. In a static object, the information is mostly redundant. We had developed an algorithm that summarizes the optimal information of each single image to generate a new sharp image of the investigated region.

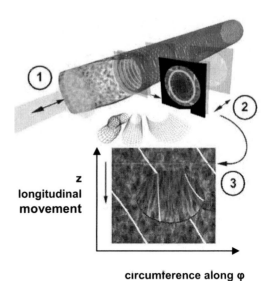

Fig. 3. The moving endoscope (1) deliverers a sequence of 2-dimensional images that are converted (2) into a 2-dimensional image with cylindrical coordinates (3) where the movement direction (z) is on one axis and the circumference correlative to the angle (φ) on the other axis.

Figure 3 shows the principle of the system. In a first step, the video stream from the moving endoscope (1) has to be digitized. From each of the 25 frames per second a set of pixels along a circle (2) is stored. Each set of pixels will then be transformed into a cylindrical coordinate system. During this operation, the circle is unfolded and results in as single line (3) of a 2-dimensional image. The circumference corresponding to the angle in the circle is on one axis. The other axis corresponds with the current position on the z-axis of the endoscope in the vessel. Throughout the movement of the endoscope along the z-axis of the vessel, the inner surface will be scanned in the described way. Each frame of the video stream results a line and according to the z-position of the endoscope, the lines are combined into the final image.

We call the system *"round-scan-endoscope"* as it scans the inner surface of the vessel while it is moved along the investigated region. In this way, a single image is generated out of the video stream, showing a 2-dimensional mapping of the inner surface. The acquired image looks similar to a dissected vessel and helps the surgeon to evaluate the anastomosis.

Development System Using A Motion - Simulation-Model

The interpretation of the digital video signal is done with the „Montivision Development Kit", which is a software that delivers a modular system for digital image processing in real time. The main task for the image-processing algorithm is to cut out and to assemble the corresponding image content to a new picture.

For evaluation purposes during the development of the algorithm we used computer simulated video data of the movement of the endoscope through a blood vessel. This was done mainly to compute comprehensible and clear data, because the image sequences retrieved from the endoscope where of too low quality for the evaluation processes during the development.

Fig. 4 shows a schematic structure of the motion simulation model. We created some 3D models of a vessel with several kinds of anastomosis (end-to-end, end-to-side, 45°, etc.) and geometric variations of the tube structure (straight/bended, round/oval) and surface (flat/bumpy). These models where used to render video streams that show a simulated view of the endoscope camera moving along the vessel. The motion of the camera was simulated with several speeds and accelerations, lateral offset, rotations, optical parameters and lightning conditions.

Fig. 3 shows a resulting image of a computer simulation after processing the computed video of the simulation with the round-scan algorithm. The white stripes in the resulting image of an end-to-side model in Fig. 3 are a test pattern. We used different kinds of test patterns that where laid on the surface of the model to see if the algorithm computes the video data back to what the test pattern should look like.

In our first 3D-computer simulations, we used a model with several idealisations concerning mainly the position and the consistent, straight movement of the camera. Further, these ideal conditions shall be adapted progressively to match real conditions during an examination of a human guided endoscope. Among other things unsteady and jerky movements, sloping viewpoints and non-axial camera positions have to be considered.

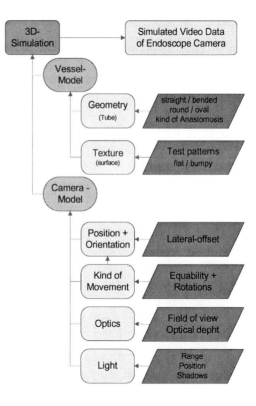

Fig. 4. Schematic structure of the motion simulation model. The view of a camera passing through a vessel is rendered to a video stream.

Technical parameters

The information that is received from the endoscopic video source is limited to 500*500 pixels, due to the round image and a frame rate of 25 images per second.

Within a vessel of 1 mm in diameter, the circumference is 3.14 mm, which is displayed within resolution of 0.0042mm per pixel. The axial resolution depends on the speed the camera is moved through the vessel and on the used frame rate.

To acquire an image of an anastomosis the region of interest (ROI) is of a few millimetres length. The endoscope should be moved approximately with 0.1mm per second. Practical tests have shown that it is very hard to move the endoscope that slowly. Therefore, each frame of the video source delivers not only one single line in the new picture but 5 to 10 lines, depending on the speed of the endoscope.

RESULTS

With the described roundscan-system, we obtained very promising results within the investigation of anastomosis on silastic vessels. Figure 5 shows results that were achieved using a mini endoscope with its diameter close to the investigated vessel. The unfolded inner surface of the artificial vessel can be seen. The stitches of the anastomosis, the 25µm thick threads, their distance and length and can be made out clearly. We used a 1.5 mm flexible endoscope from Aesculap in a 1.7mm thick artificial vessel for this examination.

The best results were archived by inserting the endoscope deeply into the lumen and gently pulling it back, passing by the anastomosis.

As we tried the system on real vessels like porcine coronary arteries we had to deal with the problem of remaining liquids that where disturbing the optical system.

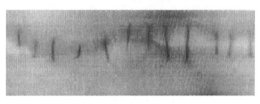
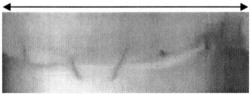

Fig. 5. Resulting images of anastomosis on artificial vessels. Clear to see are the stitches of the anastomosis, the 25µm thick threads, their distance and length. The upper image shows a correct anastomosis while the lower image reveals errors of the anastomosis.

CONCLUSIONS AND FURTHER RESEARCH

The endoluminal control of anastomosis using miniature endoscopy – roundscan technology in the porcine coronary artery training model further improves evaluation of microsurgical anastomosis. Thus, microsurgical training can be designed more effectively.

We developed the round-scan system by the use of an idealised computer simulation of an investigation.

We plan the further development of the system to make it possible to obtain images of the inner surface of the vessels of sufficient quality with existing endoscopes in microsurgical settings. The smallest endoscope that was available from Aesculap was of about 0.8mm in diameter, which made it possible to insert it into the vessel through a vain canula.

Additional adaptations of the simulation, especially varying motion speeds would help to implement a motion detection algorithm into the image processing. This would help stabilize the correct axial reconstruction of the surface. By now, this could only be realized through a steady movement of the endoscope by the surgeon. Because the investigated region is of a few millimetres length, it is possible to neglect possible rotations of the endoscope.

As a solution to deal with the liquid that is disturbing the optics of the endoscope it should be investigated if its possible to insert the endoscope with a thin, transparent coating tube, that remains in its position in the vessel, while the endoscope is moved along the vessel. With this method, it could also be possible to deal with lateral movements to guarantee the centred position of the endoscope and a straight positioning of the vessel.

3 ACKNOWLEDGEMENTS

We would like to thank the country of Upper Austria for the funding of the project, the AKH-Linz for the cooperation and Aesculap for providing the miniature endoscopes and cameras.

4 REFERENCES

[1] Schoelly W. (2003), *Microendoscope*, Pat. No. WO 03/098315.

[2] Boppart, S. A., Deutsch, T. F., Rattner, D. W. "*Optical imaging technology in minimally invasive surgery. Current status and future directions.*" Surg.Endosc. 13.7 (1999): 718-22.

[3] Fano M., Polák M. (2000) *The 3D – Object Reconstruction from 2D Slices* – Image Preprocessing CESCG 2000.

[4] Schoeffl, H., et al. "*Pulsatile Perfused Porcine Coronary Arteries for Microvascular Training.*" Ann.Plast.Surg. 57.2 (2006): 213-16.

[5] Smith W. E., Nimish V., Maislin S. A. "*Correction of Distortion in Endoscope Images*", IEEE Transactions on Medical Imaging Vol 11, No1 March 1992: 117-122.

[6] Haneishi H., Yagihashi Y., Miyake Y. "*A new Method for Distortion Correction of Endoscope Images*", IEEE Transactions on Medical Imaging Vol 14, No 3 September 1995: 117-122.

Controlling Virtual Environments by Thoughts

Christoph Guger[a*], *Robert Leeb*[b], *Doron Friedman*[c], *Vinoba Vinayagamoorthy*[c], *Angus Antely*[c], *Günter Edlinger*[a], *Mel Slater*[c]

[a]*g.tec medical engineering GmbH/Guger Technologies OEG, Herbersteinstrasse 60, 8020 Graz, AUSTRIA*
[b]*Laboratory of Brain-Computer Interface, Graz University of Technology, Inffeldgasse 16a, 8010 Graz, AUSTRIA*
[c]*Department of Computer Science, University College London, Gower Street, London, United Kingdom*

ABSTRACT

A brain-computer interface is a new device that picks up brain activity to control a device. This work shows the fascinating usage of a BCI in a Virtual Environment for navigation.
Contact: guger@gtec.at

1 INTRODUCTION

A brain-computer interface (BCI) is a new communication channel between the human brain and a computer. BCIs have been developed during the last years for people with severe disabilities to improve their quality of life [Wolpaw 1991, Pfurtscheller 1998]. Applications of BCI systems comprise the restoration of movements, communication and environmental control. However, recently BCI applications have been also used in different research areas e.g. in the field of virtual reality [Pfurtscheller 2006].

A BCI uses either slow cortical potentials, evoked potentials or oscillatory components for the control. It is well known from the literature that during a specific movement an event-related desynchronization (ERD) at a specific brain location occurs and that after the movement an event-related synchronization (ERS) occurs. The same effect can be found if only the imagination of a movement is performed.

Fig. 1. Left column – ERD during a right hand movement imagination over C3, Right column – ERD during a left hand movement imagination over C4. The amplitude attenuation (ERD) in the alpha band (10-12 Hz) is indicated in dark.

The imagination of a foot movement causes an ERD over electrode position Cz of the international 10/20 system, a right hand movement imagination causes an ERD contra-lateral over C3 and a left hand imagination over C4 (see Figure 1). Therefore the electrodes are assembled exactly over these positions in order to pick up the activity.

2 EXPERIMENT AND RESULTS

For the BCI experiments 2 bipolar EEG derivations where mounted on the subject's head (electrode positions C3 and Cz). The electrodes were connected to a portable amplifier and digitization unit (g.MOBIlab, g.tec medical engieering GmbH, Austria). The g.MOBIlab samples the data with 16 Bit and 256 Hz. Then the data is sent to the Pocket PC (see Figure 2). The Pocket PC was used to control the experimental paradigm for the BCI training of the subject. During the paradigm arrows pointing downwards or to the right side of the screen were presented. The subject had to imagine a foot movement and a right hand movement depending on the direction of the arrows. This was repeated 160 times.

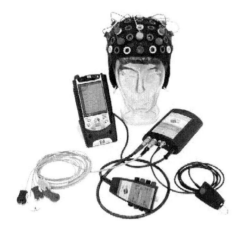

Fig. 2. Components of the BCI system. EEG cap with electrode positions according to the international 10/20 system; EEG amplifier and acquisition unit consisting of g.MOBIlab and the Pocket PC; connector boxes to plug in the EEG electrodes.

Then the EEG data was analyzed in order to distinguish the 2 different imaginations. Therefore the bandpower in the alpha band (10-12 Hz) and in the beta band (16-24 Hz) was calculated of each channel. Then a linear discriminant analysis (LDA) was used to distinguish the right hand movement from the foot movement imagination. This yields to a subject specific weight vector which can be used for the on-line experiments to control a cursor on the screen or to control a VR system [Guger 2001].

After the initial training with cursor control three subjects all with classification accuracy above 80 % were participating in an experiment in a highly-immersive CAVE like, virtual reality (VR) system (see Figure 3). The CAVE system (TRIMENSION ReaCTor) has 3 back projected screens which are 3 m by 2.2 m in size. The floor is projected by a projector mounted in the ceiling. The 3D effect is produced with shutter glasses.

Fig. 3. CAVE system which creates a 3D Virtual World.

We have used a virtual street populated by 16 avatars and shops on both sides of the street. The BCI output signal was transmitted to the VR system in order to navigate in the VE. The goal was to reach the end of the street. The subject was instructed by an acoustic cue to imagine a foot movement (double beep) or a right hand movement (single beep). If the foot movement was classified correctly the subject was moving forward, otherwise the subject was remaining on the same position (see Figure 4). If a right hand movement was correctly detected the subject was also remaining on the same position otherwise as a punishment the subject was moving backwards. Therefore only with a 100 % BCI classification accuracy the subject was able to reach the end of the street.

		subject imagined	
		foot movement	right hand movement
cue class	foot movement	forward	stop
	right hand movement	backward	stop

Fig. 4. Task to explore the Virtual World.

The accuracy was determined as achieved cumulative mileage and measured how far the subject could move. S1 had a performance of 63.6 %, S2 of 78.9 % and S3 of 85.4 %.

Figure 5 shows the ERD/ERS calculated during the training phase (top) and during the experiments in the CAVE (bottom). In the training phase mainly an ERD in the alpha and beta band is responsible for the classification accuracy while in the CAVE experiments mainly an ERS in the alpha and beta band is used for the control. The figure demonstrates the brain plasticity which is depending on the training stage but also on the way how the feedback is created and on the motivation of the subjects participating in the study.

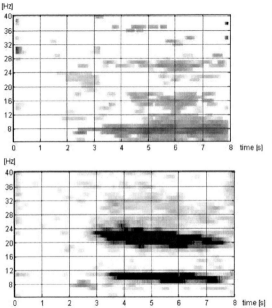

Fig. 5. ERD/ERS time curves during the training phase with the cursor (top) and the experiment in the cave (bottom).

3 DISCUSSION

The work showed that motor imagery can be used as input signal for a BCI system to control a VE in a highly immersive CAVE system. Subjects reported about an exciting experience of moving forward and backward just by the imagination of different types of movements.

Furthermore the study showed the importance of periodic updates of the subject specific weight vector. After the training phase the weight vector is trained on specific ERD/ERD patterns for a specific feedback scenario which might differ completely from the next experiments performed. Such updates are dependent on the training stage of the subject, the way of feedback, motivation and electrode positions.

4 ACKNOWLEDGEMENTS

This work is partly funded by the EC project PRESENCCIA and FFF.

5 REFERENCES

[1] Wolpaw JR, McFarland DJ, Neat GW, Forneris CA. *An EEG-based brain-computer interface for cursor control.* Electroenceph. Clin. Neurophy., 1991; 78: 252-259.

[2] Pfurtscheller G, Neuper C, Schlögl A, Lugger K. *Separability of EEG signals recorded during right and left motor imagery using adaptive autoregressive parameters.* IEEE Trans. Rehab. Eng., 1998; 6: 316-325.

[3] Guger C, Schlögl A, Neuper C, Walterspacher D, Strein T, Pfurtscheller G. *Rapid prototyping of an EEG-based brain-computer interface (BCI).* IEEE Trans. Rehab. Eng., 2001; 9(1): 49-58.

[4] Pfurtscheller G, Leeb R, Keinrath C, Friedman D, Neuper C, Guger C, Slater M. *Walking from thought.* Brain Research, 2006; 1071(1): 145-152.

Image Registration of MR/CT and Low-Count SPECT

Werner Backfrieder[a*], Wilma Maschek[b]

[a] Department of Software Engineering in Medicine, Upper Austrian University of Applied Sciences, Hauptstraße 117, 4232 Hagenberg, AUSTRIA
[b] Institute of Nuclear Medicine and Endocrinology, General Hospital Linz, Krankenhausstraße 9, 4020 Linz, AUSTRIA

ABSTRACT

Evaluating dynamics of specific radioactive tracers is a robust and accurate method for in-vivo diagnostics of tumor diseases and human metabolism. Image data provide functional information instead of anatomical details, and furthermore suffer from substantial image noise inherent to low count rates. A novel, data adaptive filtering method based on principal components analysis was developed for optimized separation of signal and noise. Enhanced images are subject to image registration with CT/MR data providing both anatomical and functional information for differential diagnosis.

Results: Accuracy of the method was tested with low count rate clinical data from betaCIT studies. Noise was sufficiently removed from image data. Registration provides encouraging results.

Keywords: emission tomography, iterative reconstruction, principal components, adaptive filtering, registration, chamfer matching

Contact: Werner.Backfrieder@fh-hagenberg.at

1 INTRODUCTION

During the last decade modern imaging methods substantially increased accuracy and specivity of diagnostic findings. High resolution computed tomography (CT) and magnetic resonance imaging (MRI) provide detailed images of anatomy and pathologies. MRI and CT enable insight into smallest structures with different contrasts, but suffer in many cases from low functional information.

Nuclear medicine utilizes tracer kinetics for diagnosis. A radioactive marker is applied directly or coupled onto biological tracer molecules. Tracer accumulation either indicates location and size of a tumor or visualizes metabolism. Typical examples are tumor staging by assessment of standardized uptake values (SUV), evaluation of image-time series for evaluation of kidney dynamics, or perfusion studies after stroke or myocardial infarction.

Tomographic imaging modalities are employed in most nuclear medicine studies. Single photon emission computed tomography (SPECT) is a versatile and robust method to acquire a series of transversal slices of the investigated body region. Spatial resolution and contrast in SPECT slices are about an order of magnitude worse compared to MRI or CT, due to the following reasons: Signal-to-noise ratio (SNR) is the major quality measure in medical imaging. SNR is modeled by a Poisson-counting process, if the number of registered counts is high, the relative error is small. Achieving a high number of counts is subject to a trade-off between patient-dose and acquisition time. Both should be kept low. Pixel-size is a further criterion, small pixels guarantee high spatial resolution at the cost of lower SNR. Besides these components controlled by the acquisition protocol, biochemical mechanisms influence image quality. The ratio of specific and non-specific bindings of tracer molecules directly manifests in the strength the useful signal. In adaptive filtering the signal is separated from noise and background signal.

Since SPECT and CT/MRI provide complementary information multi-modal imaging is desirable. In most cases SPECT image quality is too poor for application of standard registration techniques. We have developed novel filtering and reconstruction methods for signal enhancement in SPECT data. The processing pipeline comprises PCA-filtering, fully 3D iterative reconstruction, segmentation and surface based image registration.

2 METHODS

The goal of this approach is to improve SNR of SPECT data with poor statistics for further use in automated registration. The methods are described in detail below.

2.1 Principal Components Filtering

Principal components filtering (PCF) is a data adaptive filter (Backfrieder et al., 2005). In a data driven way true image structures, i.e. the signal, are separated from stochastic contributions, i.e. noise. In the first step of the filtering procedure the image is subdivided into squared regions. The data model for each region is

$$x_i = s_i + n_i , \quad (1)$$

where each tile x_i is the sum of the true signal s_i and inherent noise n_i. Furthermore the pixel j of image region i is a weighted sum of its m principal components

$$x_{ij} = \sum_{k=1}^{m} pci_{ik} \cdot a_{kj} . \quad (2)$$

The vector $\vec{a}_j$, i.e. the principal component, is collinear to the Eigenvector of the j-th Eigenvalue of the correlation matrix calculated from image regions. The image structure pci is associated with a specific PC and fulfills Eqn. 2. Summing over all m orthogonal PCs exactly reconstructs the measured image x. Sorting PCs according to descending Eigenvalues allows the following interpretation: the first PC explains the mean of the image, the second adds variations orthogonal to this mean, so does any further PC until just noise is added to the image. By the way the signal is separated from noise by orthogonal decomposition using PCs. In each image region

the signal is represented by p principal components. The separation of signal and noise is conceptually illustrated in Fig. 1. The number of PCs is different for each region and is controlled by a CHI2 test. This method is superior to linear filters with constant kernel, while it de-noises images but keepes edges.

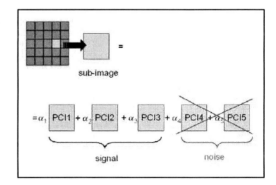

Fig. 1. Concept of PC-filtering.

2.2 Fully 3D Iterative Reconstruction

In iterative reconstruction a discrete imaging model is used

$$y_i = \sum_j a_{ij} \cdot x_j , \quad (3)$$

where each measured projection value y is the weighted sum of all pixels x, where the weights a_{ij} define the pixel-specific contributions. This model allows flexible integration of scanner geometry, scatter, attenuation, and detector efficacy into the reconstruction process. The system of equations in Eqn. 3 is solved iteratively by the ML-EM algorithm, implementing Poisson-counting statistics (Shepp and Vardi, 1982)

$$x_j^{(n+1)} = x_j^{(n)} \cdot \sum_i a_{ij} \frac{y_i}{\sum_{j'} a_{ij'} x_{j'}} \quad (4)$$

In current clinical applications projection values and pixels are grouped for each slice separately, this partitioning does not reflect physical reality.

There is a rather strong correlation between slices caused by scatter and detector geometry. We developed a fully 3D variant of the algorithm for accurate reconstruction of SPECT volume images (Backfrieder, 2003). In a 3D case, e.g. a 128x128x128 image cube, the system matrix A consists of $4*10^{12}$ elements, needing 32 Terabytes of memory, which is above the resources of even dedicated computing centers. Considering symmetries and exploiting the sparse nature of the matrix, the number of non-redundant element may be reduced to an order of 10^7, which is manageable by dedicated clusters. An ordered subsets, accelerated algorithm was implemented on an SMP cluster as a Grid application using the GEMSS framework (Backfrieder et al., 2005a). Where a Grid client-server architecture controls the reconstruction algorithm on the dedicated cluster.

2.3 Registration

For registration of CT/MRI and SPECT data chamfer matching, a surface to points algorithm is used. The algorithm employs rigid body transformation, defining 9 degrees of freedom for scaling, rotation, and translation. The registration procedure comprises the following steps:

(1) segmentation of corresponding surfaces in both image volumes

(2) generation of a distance map for the base surface using the chamfer surface-to-distance transform

(3) generation of reference-points positioned on the matching surface

(4) definition of a cost-function as distance measure of surfaces, using the distance map and the reference points

(5) assessment of the transform matrix by optimizing the cost-function using a steepest gradient algorithm in a multi-scale approach

(6) transformation of the matching volume onto the base volume and display of overlaid volumes

This method provides robust and accurate registration of complementary volume images. User interaction is reduced to segmentation of surfaces. Since PC-filtering and iterative reconstruction substantially reduces artifacts in SPECT images, the brain surface is a reliable feature for registration. The brain is segmented by a simple threshholding operation on SPECT data. In CT/MRI the brain is segmented using binary mathematical morphology, serialized by simple processing scripts.

3 DATA

Image data were selected from a clinical study providing an extremely low count-rate. In Nuclear Medicine the stage of Parkinon's disease is assessed by measuring the concentration of betaCIT in basal ganglia in relation to surrounding brain tissue regions. The binding specifity of betaCIT is very low, thus a very weak signal from ganglia is obtained. The acquisition parameters of the SPECT protocol are: 128x128 image matrix, 298 mm FOV, 2.33 mm slice thickness, LEHR-PAR collimator, acquisition time 30s / projection, and rotational increment 3 degrees, 185 MBq I123 were applied. A projection image is shown in Fig. 2.

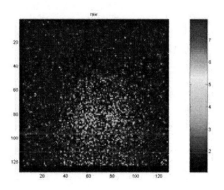

Fig. 2. Projection image of betaCIT study, acquisition time 30s, maximum pixel count 7.

MRI data were acquired on a 334x386 image matrix, FOV 232mm x 200mm, 6mm slice thickness using T2 weighting.

4 RESULTS

Accuracy is shown for PC-filtering, fully 3D reconstruction, and registration of processed SPECT images with T2 weighted MRI.

PC-filtering with confidence level 0.95 is compared to standard low-pass Gaussian filtering, kernelsize 9x9 pixels, and σ=2 pixels. Results are shown in Fig. 3. Preservation of edges and smoothing of noise by PC-filtering is clearly shown, whereas Gaussian smoothing substantially flattens edges. Even small structures are identified as signal, e.g. the two points below the basal ganglia, being classified as outliers by other filtering methods.

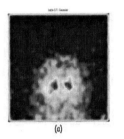 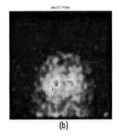

Fig. 3. Filtered projection data, (a) 9x9 Gaussian mask, (b) PC-filter with 0.95 significance

Results of fully 3D iterative reconstruction are shown in Fig 4.a. A transaxial slice through basal ganglia is shown. Tracer accumulation in the nucleus caudatus and putamen is strongly manifested and clearly differentiated from nonspecific bindings in background. Iterative reconstruction provides sharp contours, together with strong reduction of noise. SNR ratio is substantially better than with FBP, see Fig. 4.b. Beam artifacts are completely suppressed.

Registration of SPECT and MRI-t2 volumes is show in Fig. 5. Results of chamfer matching, using the brain surface as a matching feature, are visualized in a three panel display. At the left side of the figure the base volume is displayed, at

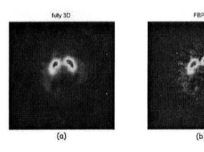

Fig. 4. Reconstruction of basal ganglia from PC filtered data. Fully 3D ML-EM reconstruction (a) and standard Filtered Back-Projection with ramp filter (b).

the right side the registered secondary volume is shown, in the center the overlay of both volumes is visualized using 1:1 intensity weighting of both modalities. The rows of the figure show the main cutting planes, transversal, sagittal, and coronal. Accurate registration is achieved, basal ganglia are correctly localized in the MR images.

5 DISCUSSION

Anatomy in addition to functional information of SPECT images potentially increases diagnostic options, e.g. accurate detection if a tumor has penetrated an anatomical barrier. Since there have been developed many algorithms for registration of MR and CT images (Maintz and Viergever, 1998), a major problem registering SPECT images is poor counting statistics and as a consequence, weak manifestation of features. The newly developed SPECT data processing strategy of PC-filtering and fully 3D iterative reconstruction results in substantial signal enhancement suitable for basic automated segmentation techniques, thus reliable features are provided. Benefits from automated registration are: results are reproducible and more objective than operator controlled manual registration.

The method is applicable to both CT and SPECT images, depending on landmarks suitable for registration. Hybrid cameras provide inherent registration of SPECT and CT within one modality, covering many clinical applications. But, there are still many dedicated SPECT systems

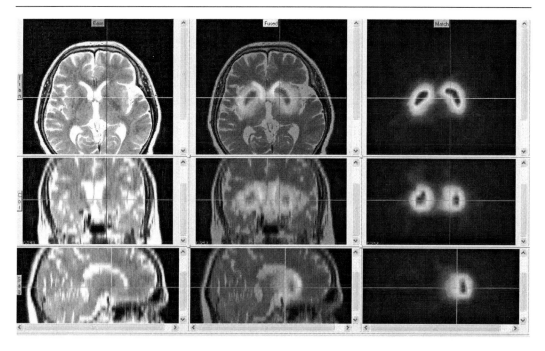

Fig. 5. Three panel display of registered SPECT and MRI-t2 volume images of a betaCIT study. Columns show from left to right native MR images, matched data, and registered SPECT data. Rows show radiological main sections of image volumes, transversal, coronal, and sagittal. Tracer accumulation is localized anatomically correct. Data may be analyzed by a linked cursor.

installed, or the need of advanced CT protocols, not implemented in hybrid cameras, require the application of the proposed registration method. Combined SPECT and MR scanners will not be available in near future, thus there is a need of dedicated registration algorithms for these modalities.

A novel algorithm for optimized signal enhancement in SPECT imaging was developed showing a high potential in multi-modal imaging.

6 ACKNOWLEDGEMENTS

Authors want to thank Prof. Dr. Martin Šámal, from the Institute of Nuclear Medicine, First Faculty of Medicine, Charles University Prague, Chech Republic for valuable discussions on PC-filtering.

7 REFERENCES

[1] Backfrieder, W., Hatzl-Griesenhofer, M., Huber, H., Maschek, W. (2005) *Multimodal image registration for low count SPECT images.* EJNM vol 32, S1, September 2005, p.76.

[2] Shepp, L., and Vardi, Y. (1982) *Maximum likelihood reconstruction for emission tomography.* IEEE TAMI, vol 1, no 2, pp. 113-122.

[3] Backfrieder, W. and Forster, M. (2003) *Locally variant VOR in fully 3D SPECT within a service oriented environment.* Procs. of METMBS'03, Las Vegas, June 23-26 2003, CSREA Press, ISBN 1-932415-04-1, pp. 217-21.

[4] Backfrieder, W., Forster, M., Engelbrecht, G., Benkner, S., Maschek, W. (2005a) *Fully 3D SPECT reconstruction in a Grid architecture.* EJNM vol 32, S1, September 2005, p.103.

[5] Maintz, A., and Viergever, M. (1998) *A survey of medical image registration. Medical Image Analysis.* vol 6, no 1, pp1–36.

RAPS: A Rapid Prototyping Framework Based on RCP, ITK and VTK

Franz Pfeifer[a*], *Roland Swoboda*[a], *Gerald Zwettler*[a], *Werner Backfrieder*[a]

[a]Upper Austrian University of Applied Sciences – Research and Development Competence Center, Hauptstraße 117, A-4232 Hagenberg, AUSTRIA

ABSTRACT

Motivation: Traditional surgery planning is often based on images originating from two-dimensional radiography. To enable surgeons to plan surgeries based on three dimensional computer models, computed tomography plays an important role. Furthermore, special fabrication systems enable the transformation of these three-dimensional computer models into real, solid models in high quality. Special applications which are able to perform filtering and segmentation of volumetric image data are needed to produce these models. To ensure user acceptance, this applications must be intuitive and high-performance. Another requirement of such systems is the accuracy of the produced models to ensure robust and accurate results.

Results: At present, a prototype application offering visualization, basic and advanced image filtering and segmentation operations is preserved.

Availability: The implemented prototype is not available for public.

Contact: franz.pfeifer@fh-hagenberg.at

1 INTRODUCTION

Rapid Prototyping is a term which has been mainly affected by the industry in the past. Today, however, rapid prototyping gains more and more in importance in medicine as well. Especially surgery planning is a promising field of application for this technique. Traditional surgery planning is based on images originating from two dimensional radiography and thus limiting the surgeons' scope (Everton L., 2004). By adding a third dimension the quality of the planning process increases dramatically. This third dimension is typically made available by *computed tomography* (CT).

In the course of the *Multi-Slice Computer Tomography* (MSCT) project, a Siemens Somatom Sensation Cardiac 64 multi-slice CT has been acquired in corporation with the General Hospital of Linz. This device enables the acquisition of three-dimensional volumetric image data in excellent temporary and spatial resolution. The aim of MSCT project is to use the advantages of the CT technology not only for rapid prototyping, but also for liver tumor diagnostics and evaluation of the cardiovascular system.

The aim of the *Rapid Prototyping in Surgery* (RAPS) project, as part of the MSCT project, is to produce solid models based on the volumetric image data acquired during CT examinations. To produce such models special rapid prototyping systems are needed. Currently, there are many different prototyping systems available, all of them differ in the procedure and materials to generate output. The most common ones are *Selective Laser Sintering* (SLS), *Stereo Lithography* (SLA), *3D milling* and *3D printing*. SLS uses a laser beam to fuse powdered materials whereas SLA uses a laser beam to harden layers of epoxy resin. 3D printing on the other hand is a combination of conventional inkjet technology and SLA, but instead of ink, 3D printers use liquid treacle and powdered plaster instead of epoxy resin. Compared to SLA and SLS 3D printing is considerably cheap. The produced models, however, have to undergo a special post processing treatment to ensure solid surfaces. Not much attention has been paid to 3D milling in medical applications since this technique is not able to produce cavities (V. E. Beal, 2004).

2 RAPID PROTOTYPING PROCESS

In this work we define the rapid prototyping process as the action of converting a series of digital images into a solid model. The workflow of this process is illustrated in **Fig. 1**. The data acquisition which is performed by a CT scan marks the start of this process. This scan generates volumetric image data which is stored as a series of two-dimensional images into a *Picture Archiving and Communication System* (PACS) using the *Digital Imaging and Communications in Medicine* (DICOM) storage format. An advantage of the DICOM format is that patient specific information as well as recording information is stored in a header of each file, allowing the assignment of a series of images to a patient. As several CT scans are performed, more than one series of images are assigned to one patient. The import module separates the different series by parsing the header of each DICOM slice. Another task of this module is to generate a volume based on the DICOM images of a certain series.

This volume is presented to the user who may perform filtering and segmentation operations to extract and enhance a *Volume of Interest* (VOI) as input for subsequent processing steps.

A common image data manipulation process starts with cropping the volume to the size of the VOI. This not only reduces required memory but also leads to faster filter execution. Later inadequacies of the imaging process like background noise and partial volume effect are reduced by the use of smoothing filters. After these preprocessing steps the VOI is segmented at the use of simple methods like region growing or sophisticated deformable contours, depending on the anatomy to segment and its contrast to the background.

One of the most important steps in the rapid prototyping process is the generation of a surface model based on the extracted VOI. To ensure user acceptance, this task is fully automated.

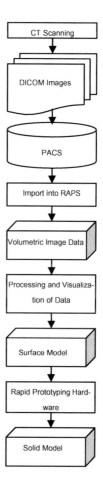

Fig. 1. Workflow of RAPS.

This surface model may undergo further user interactions like clipping, translating, rotating or merging with another surface model. After the user has finished model manipulation, the surface model has to be exported into a file format the rapid prototyping system is able to read; the *Virtual Reality Model Language* (VRML) and

Stereo Lithography (STL) are the most commonly used file formats.

3 ARCHITECTURE

One of the main goals of the MSCT project is to develop a software platform with intuitive user interface and that is easily extendible by other programmers. Therefore exact differentiations between the functional aspects as well as the presentational aspects of the platform have to be made. Furthermore, clear definitions of interfaces are essential to allow other programmers to extend or to adopt the platform to their needs. An important aspect of the platform is to ensure high visualization and computational performance. To achieve this aim, the functional layer is entirely based on C++ libraries. The main goal of the presentation layer, on the other hand, is to provide an easy to understand, highly flexible and extendible Graphical User Interface (GUI). To satisfy this requirement, the whole presentation layer is implemented in Java. Nevertheless, this Java GUI can be substituted by a C++ GUI without any problems.

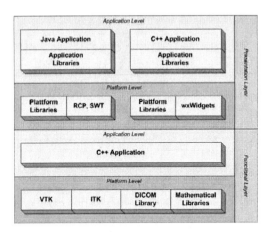

Fig. 2. HIP-Image Processing Platform.

The next sections describe the architecture of the functional layer as well as the presentation layer in more detail. **Fig. 2** shows the top-level delineation of the platform which is called *Hagenberg Imaging Platform* (HIP). As illustrated, the presentation layer as well as the functional layer are separated into an application level and a platform level. This separation of concerns ensures that changes to the application do not affect the platform itself, but changes to the platform are leading to changes in the application.

3.1   Functional Layer

At present, the functional layer consists of two parts: one part performs visualization of volumetric image data, the other part is responsible for loading DICOM images, filtering, registration and segmentation of volumetric image data. Both parts are based on open source frameworks implemented in C++. While visualization is performed by the *Visualization Toolkit* (VTK), filtering, registration, segmentation and image loading is performed by the *Insight Segmentation and Registration Toolkit* (ITK). The two frameworks are distributed by Kitware (www.kitware.com) and are described in more detail below. The DICOM library and mathematical libraries shown in **Fig. 2** are currently not specified yet.

3.1.1 Visualization Toolkit: As mentioned above, VTK is an open source project enabling users to quickly visualize data without knowing any details of the underlying 3D graphics and modeling library. On Microsoft Windows Systems this is most commonly OpenGL (Wright R., 2000). Furthermore, VTK follows the object-oriented paradigm, allowing a reusable system design. A noticeable feature of VTK is the so called visualization pipeline which consists of (a) objects representing data (data objects), (b) objects operating on data (process objects) and (c) a connection of dataflow between these objects (Schroeder W., 1996). The purpose of data objects is to store information. They provide methods to create, to access and to delete information. Process objects, in contrast, provide methods to manipulate information stored in data objects. Therefore, process objects provide at least one input connection and one output connection. Depending on the task of a process object there may be more than one input and output connection. Process objects can be connected in such a way that the output of one process object is the input of another process object. This procedure builds a so called visualization network. To keep the output of a visualization network up-to-date, each process object determines if its parameters have changed or if its input has changed. If so, the process object executes again, otherwise no action is performed. The advantage of this approach is to avoid unnecessary executions, leading to a higher performance of visualization, see **Fig. 3**. Another important feature of VTK are wrapper layers, allowing VTK to communicate with other programming languages such as Java, Python and Tcl. As mentioned earlier, this approach enables high performance and flexible application design. Moreover, VTK supports distributed parallel processing using *Message Passing Interface* (MPI) and hardware volume rendering based on *VolumePro*.

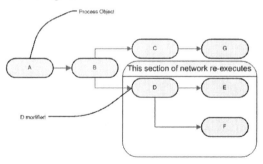

Fig. 3. VTK pipeline execution model.

3.1.2 Insight Segmentation and Registration Toolkit: The ITK library covers the main tasks of image processing. Besides basic linear filters and methods for image manipulation also novel and comprehensive concepts like deformable image registration and level sets for image segmentation are provided. In addition to the main focus of ITK, image registration and segmentation, algorithms for statistical analysis of input image data and fuzzy logic are part of this library. All of these functionalities are assembled from small building blocks like metrics, optimizer or interpolation classes based on object oriented programming. This enables programmers to reuse these building blocks to form new functionality. Similar to VTK, the insight toolkit uses a pipeline concept to forward results when multiple filters or I/O components are executed. A fundamental paradigm of ITK is applicability on input data of arbitrary dimension. Most of the filters follow this paradigm and are absolutely autonomous from image dimension (Yoo T., 2004) (Ibanez L., 2005). Although this demonstrates a very high level of flexibility and reusability, an increase in runtime complexity has to be noted. As all image data is internally handled as one-dimensional array, random voxel access is quite costly for large multi-dimensional image data. Next to data storage runtime is prolongated by object oriented overhead and overhead that results from generic design.

As a consequence some parts of ITK are extended or replaced by mask based filters and algorithms optimized for processing two- and three-dimensional image data. Due to these improvements the computational time for concepts of mathematical morphology could be reduced by a factor of 20 on the average.

So several time consuming ITK filters were replaced by faster implementations but the very valuable and sophisticated concepts like level sets and deformable registration, the strength and benefit of ITK, don't need improvement.

3.2 Presentation Layer

The task of the presentation layer is communication between the use and the functional layer by offering a GUI. Compared to C++, Java offers many freely available tools to build GUIs. Furthermore, Java is platform independent, meaning, that Java programs are able to run on every operating system, as long as a *Java Virtual Machine* (VM) is installed. C++ programs have to be recompiled in order to run on different operating systems. A problem which had to be mastered is the combination of the C++ functional layer and the Java presentation layer. As mentioned earlier, VTK offers wrapper layers which enable communication between C++ and Java;

this communication is handled by the *Java Native Interface* (JNI). JNI enables programmer to incorporate Java code with native code, written in C or C++ or vice versa (Liang S., 1999). JNI is used by the Java Platform itself in many parts, i.e. the *Abstract Windowing Toolkit* (AWT) is based on JNI.

An outstanding feature of HIP is the use of *Eclipse Rich Client Platform* (RCP) as presentation layer. Eclipse is an open source project, with the aim to provide Java tools and infrastructure for software development. Eclipse itself is well known for the Integrated Development Environment (IDE), which itself is based on a generic tooling platform. This tooling platform is in turn based on the Eclipse RCP (McAffer J., 2006).

3.2.1 Eclipse Rich Client Platform: Rich clients started in the early 1990s with the intention to replace terminal clients being state of the art. Rich clients enabled users to focus on their work and not on the system and in turn increased the user's efficiency in writing code. The characteristics of rich clients are a rich user interface and high performance due to the elimination of network traffic. Another characteristic of rich clients is that the application of reusable components shortens development time and, as a consequence, reduces development costs. Furthermore, the middleware provides frameworks and infrastructure which shorten the development time as well (McAffer J., 2006).

In addition to the above mentioned characteristics, the Eclipse RCP offers a plug-in concept which enables programmers to extend or adapt existing Eclipse RCP applications to their needs. Another big advantage of the Eclipse RCP is a special mechanism allowing an automated update of parts of an application. RCP uses the Standard Widget Toolkit (SWT). Like AWT, SWT uses native OS widget libraries which lead to an increased performance of the GUI. Besides SWT, JFace, a GUI framework based on SWT, offers components like dialogs, wizards and viewers, is integrated into RCP, enabling programmers to rapidly build complex user interfaces.

4 RESULTS

Based on the architecture described in section 3, a prototype application was developed. This prototype analyzes and imports a series of DICOM images and visualizes the volume in the three main sections (axial, coronal, sagittal), as presented in a usual diagnostic viewer. A three-dimensional volume is rendered in a fourth view. This volume is color- and transparency-coded, meaning that the intensity of each volume element is mapped to a certain color and transparency value.

All of the four views react on user inputs and are able to perform actions like rotation, translation and zoom. In addition, the three planes are able to change the window-level, add seed points and add contours to the view. Furthermore, the prototype offers noise suppression, region growing and threshold filters, all of them working in three dimensions.

Figures 4-7 show screenshots of the prototype application and the final plaster model. The example below shows how to use the prototype to segment bone structures from a volumetric image data.

4.1 Example: Spine Fraction

As described above, the loaded volumetric image data contains noise that has to be eliminated to allow an exact distinction between soft tissue and bone structures. In a first step a Median smoothing filter with a radius of two is applied to the volume followed by a recursive Gaussian smoothing filter to reduce the noise. The standard deviation of the Gaussian smoothing filter in x, y and z direction has been set to 1.5, 1.5 and 1.0, respectively. **Fig. 5** shows the resulting volumetric image data where soft tissue and bone structures are noticeable separated. To remove the soft tissue a threshold is applied, see **Fig. 6**; **Fig. 7** shows the printed solid model of the spine.

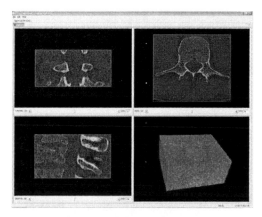

Fig. 4. The loaded DICOM volume.

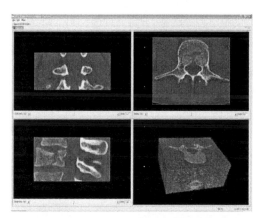

Fig. 5. Results after noise suppression.

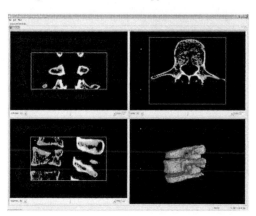

Fig. 6. Results after removing soft tissue.

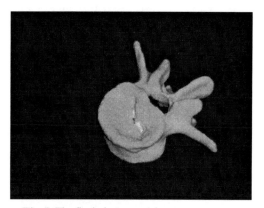

Fig. 7. The final plaster model.

4.2 Performance

Table 1 compares the runtimes of the filters used in the above described example and the runtimes of the corresponding filters of an application called *VolView*. The volume used for this example has an extent of 396x335x183, requires 46.3 megabytes of memory and took about 7 seconds to load. Both applications were executed on a Pentium 4, 3.4 GHz with 2GB of memory on Windows XP Professional.

Table 1. Runtimes of MSCT Prototype filters compared to VolView filters.

Applied Filter	RAPS Runtimes (in milliseconds)	VolView Runtimes (in milliseconds)
Gaussian Smooth	3,139	5,251
Median Filter	72,9441	119,724
Threshold Filter	625	250

5 DISCUSSION

Current results show that the applied technological approach is very promising. Particularly, the usage of Eclipse RCP in combination with ITK and VTK pushes rapid application development. Although the combination of Java and C++ forces an additional application layer the complexity does not noticeably increase. By using the Eclipse RCP, the effort of building a client application is reduced to a minimum. All important GUI components as well as the event handling are already implemented and only need to be adapted to the programmer's requirements. The Eclipse RCP is easily expandable and new features can be implemented and installed with hardly any effort. The use of VTK in combination with ITK makes it easy to implement the visualization and manipulation of DICOM images and thus frees the developer from writing complex and time consuming code.

Problems were faced with incompatible threading environments used in Java and C++. Java is a multi-threaded environment, which means that user interaction and the application are separated into different threads. This especially caused problems during rendering tasks of VTK. Nevertheless, this problem has successfully been solved by attaching the event thread to the application thread. Memory leaks posed another problem during the development of the prototype. This problem has been overcome by manually deleting the VTK Java objects instead of letting Java take control of the object references any longer.

Future development will focus on an application for oral and maxillofacial surgery. For this purpose, a CT scan of a plaster cast of a patients jaw has to be registered with CT volume of the patients head. This registration is necessary to remove artifacts originating from braces of the patient. In addition, further calibration of the rapid prototyping hardware as well as further tests of the currently used materials will be performed to ensure best possible solid models.

REFERENCES

[1] E. L. Santos da Rosa, C. F. Oleskovicz, B. Nogueira Aragao. *Rapid Prototyping in Maxillofacial Surgery and Traumatology: Case Report.* Braz Dent J 2004;15:243-247.

[2] V. E. Beal, C. H. Ahrens, P. W. Wendhausen. The Use of Stereolithography Rapid Tools in the Manufacturing of Metal Powder Injection Molding Parts. J. of the Braz. Soc. Of Mech. Sci. & Eng. 2004:XXVI, No. 1:40-46.

[3] W. Schroeder, K. Martin, B. Lorensen. *The Visualization Toolkit: An Object-Orientated Approach to 3D Graphics.* Prentice Hall, Upper Saddle River, New Jersey, 1996.

[4] L. Ibanez, W. Schroeder, L. Ng, J. Cates, Insight Software Consortium. *The ITK Software Guide.* Kitware, Inc., 2005.

[5] S. Liang. *The Java Native Interface: Programmer's Guide and Specification.* Addison-Wesley, Massachusetts, 1999.

[6] J. McAffer, J.-M. Lemieux. *Eclipse Rich Client Platform.* Addison-Wesley, Upper Saddle River, New Jersey, 2006.

[7] R. S. Wright, Jr., M. Sweet. *OpenGL Super Bible.* Waite Group Press, Indianapolis, Indiana, 2000.

[8] T. Yoo. *Insight into Images: Principles and Practice for Segmentation, Registration and Image Analysis.* A.K. Peters, 2004.

Workshop:

Logistics and Supply Chain Management

Chairman: Franz Staberhofer

Scientific Board:
Peter Klaus, Fraunhofer ATL
Bruno Krainz, MAN Nutzfahrzeuge Österreich AG
Franz Staberhofer, FH OÖ Campus Steyr
Tatjana Wallner, FH OÖ Campus Steyr

How Can Individual Procurement Processes Fit in Comparable Supply Chains?

Sabine Bäck[a], Uwe Brunner[b]

[a]FH JOANNEUM, MSc Supply Management, Werk-VI-Straße 46, A-8605 Kapfenberg, AUSTRIA

[b]Rigips Austria, Leitung Werkslogistik, Wiener Neustädterstr. 63, A-2734 Puchberg am Schneeberg, AUSTRIA

ABSTRACT

Motivation: In literature and industrial life there are various discussions about the importance of purchasing and procurement processes and their impact on the supply chain. However, based on a study of BME only 12% of companies have clear targets and only 13% of them are measuring their success. Even, if they do it, is hard to benchmark the efficiency parameters because process layouts and measurement systems are often incomparable. Furthermore, surveys are pointing out that purchasing in companies does not have the appropriate significance other than e.g. sales and distribution departments. This could also be seen in the salaries of purchasing staff and sales management staff.

Results: This paper has a focus in sorting out the differences in integration of procurement and purchasing tasks in SMC and big companies in various branches. Therefore, the responsible persons for procurement in 30 companies were interviewed on the basis of a standardized questionnaire. Interviewed companies were split into following branches: Industry, Wholesalers and Retailers and Service Providers.
The processes were modeled with the ARIS-method to have a standardized view on that topic and to develop the basis for a systematic analysis and benchmarks.

In industry SMCs purchasing is not seen as a core process, although it has an essential impact on production costs and price calculation. Thereby companies are not able to use the instruments of process- and supply chain management like SCOR because parts of supply chain are not managed.

In contrast purchasing activities of retailers (very small companies) are driven by those big suppliers with whom they interface.

Regarding service providers such as hospitals, the transparency and knowledge about effective and efficient purchasing and procurement is more or less rare because it is more budget driven.

All in all, there are various potentials to optimize purchasing and procurement activities in the different branches and fields. Particularly, it has to be a focus on the development of the knowledge of methods to see procurement and purchasing process as an embedded value within the supply chain. Precise procedures and recommendations for the implementation of the above mentioned supply chain process would be given in this article. These proposals are based on the possibilities to measure, compare, and steer the internal and external processes.

Contact: sabine.baeck@fh-joanneum.at

1 INTRODUCTION

One of the companies' strategies during the last 20 years was to move from being an all-rounder to an expert (*see. Göpfert*). Because of this outsourcing trend a new supplier profile has been created, the system provider. In response to this radical change, particularly, the purchasing department was confronted with new business methods like Lean Production, Just in time, Kanban, Integrated Materials Planning, Total Quality Management (*see. Arnolds H./Heege F./Tussing*) and Supply Chain Management. In addition, supported by the focus on process management and new IT-Technologies (e-Business, ERP-Technologies, Advanced Planning Systems) speed and efficiency of the supply based on the demand has grown.

In literature (surveys) the value of the purchasing function is recognized as one of the strategic ones with a high potential of improvement. For this reason business reference models (like SCOR, VCOR, etc.) have been developed on a highly abstracted level to be in the position to evaluate those savings.

Nevertheless, in practice, especially SMC's (Small and Middle Companies) are not following this trend yet.

2 BASIC PRINCIPLES

Frequently, the tasks of the purchase department are clustered into operative and strategic ones. A strict differentiation of the content for those two classes cannot be deduced from literature (e.g.: research of the procurement market; *Boutellier* and *Schulte* define it as a strategic task, *Bogaschewsky* as an operative one). All in all, the procurement process itself contains both.

The procurement process includes all steps from the appearance of the demand up to the fulfillment and ends with the payment of the invoice.

Based on the definition of *Hartmann and Burt et al.* the procurement process can be split into the following steps. (see: Fig. 1)

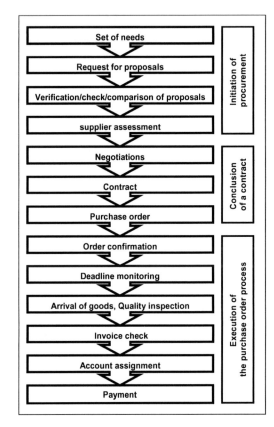

Fig. 1. Standard Purchasing Process.

The starting point for the procurement process is the identification of a particular set of needs /demand within the production or other departments of a company.

Afterwards, if necessary (e.g. no actual contract is available), a search process for suppliers who would be able to satisfy the requirements (the right quality, time and price) has to be implemented. The output is the request for proposals.

The next step is the verification, check and comparison of the proposals. Amongst others, *Kerkhoff* proposes the Multiple-Approach for global sourcing, which means that the volume of demand is allocated to different suppliers. As a consequence the company avoids the risk of dependency.

A supplier assessment for selective partners follows. It is an evaluation before an adduced performance. One or more suppliers are selected.

Negotiations are done with the potential supplier. The importance of this process step depends on the purchase volume. As mentioned before, the price, delivery schedules and defined quality is fixed now.

Nowadays when companies practise global sourcing, the contract management is very important (*see Kerkhoff*). Different cultures and countries require the knowledge of special legal requirements.

The strategic part of this process ends with the creation of a purchase order. The relationship between customer and supplier is fixed.

The following operative tasks consist of an order confirmation and a deadline monitoring. It ends with the arrival of goods and quality inspection.

At last the contract ends with the payment of the invoice. Not many companies see invoice check, account assignment and payment as a part of the purchasing department.

To meet the challenge of best practice source process management (SPM) within supply chain management, it is not sufficient to implement standard procurement processes for once (normal order, call off orders, contracts, consignment stock, vendor managed inventory, etc.), together with suppliers, companies have to continuously check and redesign also their internal and external organization, interfaces and procurement strategies (backward integration, in sourcing etc.).

3 POTENTIALS IN PURCHASING PROCESSES

3.1 The Survey

To find out the potentials of purchasing processes within the supply chain, 27 companies in the field of industry, service and retail were analyzed. One third of the companies were industrial companies, another third service oriented companies and the last third were retailers. About 25% of them were members of trade chains, which is important to note here, because a lot of them have centralized purchasing activities and decentralized procurement tasks to fulfill.

The focus of the survey was put on small and medium sized companies (SMCs), therefore about 80% were out of that field. The remaining part were bigger, mainly industrial companies, because it was the intention to have a certain possibility to compare purchasing processes of SMCs to the processes of international big players.

The variety of companies is shown on the wide range of business they do – e.g. supermarkets, haircutters, automotive, hospitals, bank institutes, insurances, etc.

Because of the different fields, the wide range of branches and business, it is important to have a structure of purchasing process types. In literature usually a lot of structuring methods can be found dealing with purchasing processes. All in all as this is not the subject we want to deal with, we have decided to categorize the purchasing processes according to *Hamm*. He differentiates between the following main purchasing processes:

- Traditional p-processes
- Market oriented p-processes
- Demand driven p-processes
- Supplier driven p-processes
- Relationship oriented p-processes

Based on the categorization of purchasing processes it is important to mention the main key questions or key points we have analyzed in our survey.

- Who is responsible for purchasing processes in companies?
- What are the main process steps used in reality?

- What are the starting and ending events of the processes?
- How intensive is the IT-support of p-processes?
- Are there any specific interfaces in the p-processes?

3.2 Consolidated Findings

Responsibility for purchasing processes:
In general it must be mentioned that a wide range of functions – here executive management, sales, administration project management and naturally purchasing or procurement staff - are doing purchasing activities in companies.

It was found out that in smallest and small companies mainly the executive managers or sales people are responsible for purchasing materials (more than 80%). In this set there was no case of small and smallest companies having a procurement or purchasing department. In comparison, every big player has installed a purchasing department. In the picture below you can see the responsibilities for purchasing processes within the different fields. In connection with the field it is very interesting to see that the industrial sector has the biggest share of purchasing staff, although the part of big players was the biggest in this survey. Retail organizations are strongly driven by the market and obviously this reflects the organization of purchasing in retail.

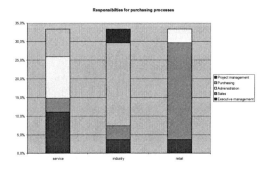

Fig. 2. Responsibilities for purchasing processes.

Process steps:
Looking at the process steps as well as on the starting and ending events in purchasing processes, two main alternatives can be seen:

- Process end event is "Goods Receiving"
- Process end event is "Payment"

The processes mainly start with a kind of demand, which has different reasons: E.g. customer inquiry, stock out situation, periodical deliveries. It is an interesting question what the end of a purchasing process should be and who should be responsible for the payment, but this is more of a question of the company's philosophy. And therefore we decided to leave this part to the companies.

In general, however, it was found out that the process steps of purchasing indeed have a wide range of variation. But looking at the activities in the process, a kind of similarity was detected. *The processes are strictly focused on traditional purchasing activities beginning with a detection of demand, followed by inquiries, offer comparison, contracting, offering, delivery, receiving and booking goods, checking the invoice and possibly payment.*

Only a few companies have a kind of strategic purchasing and especially the small and smallest companies neither have a kind of institutionalized purchasing nor do they show any strategic ideas in the field of purchasing.

Regarding the classification of purchasing processes it must be mentioned that 37% of the companies have strongly market driven purchasing processes, which means that there is a focus on the prices of the materials. The other 37% are maintaining a relationship driven purchasing process, which means that they try to integrate the supplier and to install a good partnership with them. Only 18% still have traditional purchasing processes in which the technical departments are represented as the stronger party. There is no distinct integration into the supply chain processes within the logistical processes.

IT-support:
It is not hard to anticipate that in general ERP-systems are used by big and medium sized companies. But it is interesting that nearly 30% of small and medium sized companies are using their own sector-systems and inventory control systems. And it is also predictable that the usage of Microsoft Office Packages amounts to over 35% in small and medium sized companies.

Despite the ambitions of ERP-system providers to drive forward their system in small and medium sized companies, we have to state that no company in this set is working with this kind of system yet.

Process interfaces:
It was already mentioned that the processes are still traditional purchasing processes and that there is a more or less a strong focus on the price. Certainly the impact is that there are various interfaces – organisational, IT-technical, medial – in the processes, which are potentials for process improvements.

All in all it can be said that the bigger the company the higher the number of interfaces. Additionally to that aspect we couldn't see any improvement efforts during identifying and modelling the processes.

4 RECOMMANDATIONS REFLECTED WITH LITERATURE

Based on the survey and standards discussed in literature we want to give recommendations for the integration of purchasing processes in the supply chain. Furthermore, there should be a standardized concept for controlling the supply chain with key performance indicators particularly for small and medium sized companies.

Referring to the five main questions at the beginning of this article, we want to give a guideline for improving the performance of purchasing processes.

Responsibility:
As it can be seen in the findings, purchasing activities in SMC's are done by managers of sales department or executive staff additionally to normal business. There is a lack of time for efficient purchasing processes – operational and strategic ones - as well as for training and know-how management in the field of purchasing.

Irrespective of the amount of employees, every company must have clear responsibilities for the tasks connected to purchasing and procurement. Firstly SMC's have to analyse their role in the supply network and set their objectives in connection with their key performance indicators in an appropriate way.

Purchasing Process steps:
The results of the survey show the traditional orientation of the processes and that there is a lack of integration in the supply chain. Also the survey of the University of Applied Sciences Steyr has shown that cooperation models such as VMI or CPFR which require an intensive and long term partnership within the supply chain actors have a non significant priority. Furthermore, knowledge about purchasing process steps is not as high as it should be for a clear process ownership in the supply chain.

To generate advantages of a supply network, companies have to focus their efforts on best in class defined processes and on getting a reasonable knowledge and alignment of processes instead of functions.

Also starting and ending events of operational purchasing processes must be defined globally – beginning with demand detection and ending with invoice checking and payment in order to avoid media breaks in the process and to support liquidity and cash flow management.

IT-support:
It doesn't make sense to implement huge ERP-systems in SMC's because implementation and ongoing costs are very high. But the integration in the supply network implies the necessity to use e-applications and to participate in e-business scenarios.

As mentioned before, indicators are useful for steering a company in an efficient way. Due to this condition, data storage is indispensable.

Process interfaces:
As seen in the process models of the investigated companies there are a lot of interfaces within the operational purchasing process. These interfaces must be reduced by installing process oriented structures in the companies as mentioned above. After having eliminated interfaces in the operational process, the processes must be integrated in the supply chains and the process interface there must be analysed and later optimized as well. To optimize the supply chain, different types of reference models – SCOR model on level 2 could make sense - can help companies to work on the efficiency of the supply network.

5 ACKNOWLEDGEMENTS

We want to express our appreciation and thanks to all companies which have helped us to acquire the data for this survey and also to the students of the University of Applied Sciences, Department of Industrial Management, Year 03, for their efforts and cooperation with the companies, identifying and modeling purchasing processes.

6 REFERENCES

[1] Arnolds,H. and Heege,F. and Tussig,W. (1998) *Materialwirtschaft und Einkauf,* Applied. Gabler Verlag, Wiesbaden, 359.

[2] Boutellier,R. and Locker,A. (1998) *Beschaffungslogistik*, Applied. Hanser Verlag, Wien, 40.

[3] Bogaschewsky,R. (2003) *Management und Controlling von Einkauf und Logistik,* Applied. Deutscher Betriebswirte-Verlag, Gernsbach, 28.

[4] Burt,D.N. and Dobler,D.W. and Starling,S.L. *World Class Supply Management The Key to Supply Chain Management, Applied*: 7th edition, McGraw-Hill/Irwin, Boston et al., 16.

[5] Göpfert,I. (2005) *Logistik Führungskonzeption,* Applied. Verlag Franz Vahlen GmbH, München, 236.

[6] Hamm,V. (1997) *Informationstechnik basierte Referenzprozesse,* Applied. Gabler Verlag, Wiesbaden.

[7] Hartmann,H. (2002) *Materialwirtschaft. Organisation, Planung, Durchführung, Kontrolle,* Applied. 8. Auflage, Dt. Betriebswirte-Verl., Weinheim., Gernsbach, 454.

[8] Kerkhoff,G. (2005) *Zukunftschance Global Sourcing,* Applied. WILEY Verlag GmbH & Co. KGaA, Weinheim, 49; 185-212.

[9] Schulte,G. (2001) *Material- und Logistikmanagement,* Applied. Springer Verlag, Dresden, 208.

[10] Staberhofer,F. Hrsg., (2005) *Studie zu Untersuchung der marktgerechten Logistikleistung,* Applied. FH ÖÖ Forschungs & Entwicklungs GmbH, Steyr, 6.

MCSP – MULTI CUSTOMER SUPPLIER PARK
A Solution Approach for the Automotive Supply Industry in CENTROPE/AREE

Klaus Schmitz [a]

[a] Fraunhofer Project Group for Production Management and Logistics, Theresianumgasse 27, 1040 Vienna, AUSTRIA

ABSTRACT

In 2008 within a radius of 300 km around Vienna eleven assembly plants will produce more than three million cars. The outcome of this is a business potential for suppliers of about 20 billion euros within the AREE. To participate in this boom, a local production factory close to the customers can be a key advantage for suppliers. Supplier Parks are an alternative to the high risk option of building a separate factory. A Supplier Park reduces the suppliers' overall investment and operating costs because of synergy effects.

Contact: Klaus.schmitz@fraunhofer.at

1 INTRODUCTION

The new EU-Members (Hungary, Czech Republic, Slovakia, Poland, and Slovenia) and Romania, the EU-candidate country, are called Automotive Region Eastern Europe (AREE). These countries provide an opportunity for OEMs (Original Equipment Manufacturers) and suppliers for cost reduction and market aspects on account of factor cost advantages (especially low wages) and the favourable position in Europe. One big challenge for OEMs in the AREE is to cope with enormous logistics costs. Companies source most of the parts from historically developed procurement networks in its original countries such as Germany and Korea. Consequently, actual transport operations of parts and components across the world often eliminate the labour costs' advantages within the AREE.

2 AREE – AUTOMOTIVE REGION EASTERN EUROPE

Within 300 km of Vienna, the eleven automotive plants will produce more than 3 million cars in 2008 (Sihn, Palm, Matyas, Kuhlang, 2006), which accounts for 4 percent of the annual purchasing volume of the automotive sector. By OEMs providing such a huge capacity, the suppliers have a chance to participate in the development of a procurement potential of approx. 20 billion euros in the region (Sihn, Palm, Matyas, Kuhlang 2006). Figure 1 gives an overview of all OEMs within 300 km of Vienna.

Fig. 1. OEM assembly plants in the AREE.

For the suppliers from original locations such as Germany and Korea the question arises what the best supply option for the new OEM-plants is. There are three options:

1. Serve the new factories from existing locations
2. Build a separate plant close to the customers
3. Rent a turnkey factory in a supplier park.

The first option implies high logistic costs, large stocks and high throughput times. The PSA factory in Trnava, for example, has a supply volume of 70% which is obtained from locations more than 300 km away (PSA 2006). The engines, for instance, are carried by trucks over 1300 km from the French factories in Tremery and Douvrin (Automotive News Europe 2006).

The second option means that the supplier has to take the risk to build a new factory within the AREE. Its success mainly depends on the most favourable factory size, the scale effects (manufacturing techniques used, overhead costs, etc.), the divisibility of the process stages and the customers' locations. High investment costs in land, buildings, equipment and machinery are necessary. In addition to these direct investments, high start-up costs such as employee trainings, costs for unexpected problems and high expenditure on supervision, coordination and control have to be taken into consideration.

The third option is to rent a factory in an existing OEM supplier park, e.g. in Lozorno (VW) or Trnava (PSA) or in a Multi Customer Supplier Park (MCSP). The renting of a factory in a traditional supplier park is particularly appropriate for Tier-1 suppliers with products of high variance and/or high volume. An alternative for other supplier types is to move into a MCSP which allows reductions in start-up and operating costs. Furthermore a MCSP offers the opportunity to rent a turnkey factory within a park, and thus avoiding high investment costs especially for suppliers that are unable to establish an own plant within the AREE. Moreover the MCSP provides an environment for suppliers that need a highly fixed capacity for the economical operations because of increasing amounts of clients. The concepts of Supplier Park and Multi Customer Supplier Park are explained in the following paragraphs

3 SUPPLIER PARKS

Global tier 1 suppliers, (Johnson Controls, Lear or Fauricia) settle down – more or less voluntarily - in existing supplier parks of OEMs like Lozorno (VW) or Trnava (PSA); they deliver parts with high variance and/or with high volume. A supplier park is a cluster of more than two suppliers located adjacent or close to a final assembly plant. The well-defined area includes buildings as well as infrastructure and is purpose-built in order to serve the assembly plant and the suppliers. An operator provides and maintains the whole supplier park. Objectives of parks are cost reductions and service improvements of the procurement logistics as well as protecting business relationships (Gareis 2002; Verband der Automobilindustrie 2003; Larsson 2002; Miemczyk, Howard, Graves 2004; Sako 2003)

3.1 Drivers

Since the 90's supplier parks are being developed in the European automotive industry. Its drivers are described below (see table 1) to understand this trend in more detail (Klug 2003; Westkämper, Freese, Bischoff, Barthel, Lehnert 2005).

Table 1. Supplier Park drivers.

Reasons	Description
Increasing variant diversity and shorter product life cycles	Due to the growing individual customer requirements and the saturated triade markets, the automotive sector reacted by adding new market segments (new model lines) and equipment-variants with shorter product life cycles over the past 10 years. The resulting complexity of products and processes can be reduced by supplier park.
shift to module- and sys tem-sourcing (Outsourcing)	The OEM's shift to module- and system-sourcing reduces costs and complexity, but makes it necessary to synchronize production line and supply chain (Just-in-Time) due to the multitude of variants and the high value of the parts. Storage is not possible any more (number of variants, delivery time, and flexibility). Due to its proximity, the supplier park guarantees supply reliability.

Increasing customer requirements (delivery time, change flexibility)	Short delivery times and short notice request changes of the car configuration just before the start of production need highly integrated processes with short cycle times. A supplier park can meet all these requirements.
Government grants	The provision of secure, high-grade jobs and the development of the region into an automotive production centre are drivers for government grants.
OEM's market power	The driving force for suppliers to move to the proximity of the OEM's factory is its market power.
Saving potentials in logistic costs	Due to integrated supply processes and the OEM's proximity, savings in logistic costs can be achieved (Elimination of packaging material, reduction of storage stages, etc.).

3.2 Partner model

Figure 2 shows the partners involved in a supplier park (Westkämper, Freese, Bischoff, Barthel, Lehnert 2005).

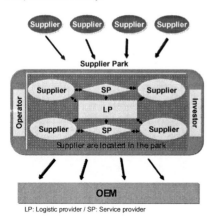

LP: Logistic provider / SP: Service provider

Fig. 2. Production locations in the AREE.

OEM: OEM factory receiving the delivered parts.
Supplier: Tier-1 suppliers within the supplier park directly supply the OEM factory with parts.
Logistic provider (LP): The Logistic provider is responsible for the logistics between the suppliers and the OEM factory (one or more logistic providers are possible).

Service provider (SP): A service provider offers non-logistic services such as canteens, fire department, garages, etc.
Operator: The operator holds the park's overall responsibility. It supervises the central services and infrastructure.
Investor: The investor finances the supplier park. Business support programmes and government grants can have a supporting function by either giving financial or infrastructure support.
Suppliers outside the park: Outside suppliers complete the network by providing the park's suppliers with necessary parts.

3.3 Characteristics

A supplier park can be characterised by the criteria listed in table 2 (Westkämper, Freese, Bischoff, Barthel, Lehnert 2005).

Table 2. Supplier Park characteristics.

Criteria	Typical specification
Location	Close proximity to customers (usually there is a distance of a few hundred meters)
Parts spectrum	High volume or high variant modules / systems
Value added ratio	Activities with low value added like module storage and sequencing, kit-building and final assembly.
Number and types of suppliers	Between 6 and 10 First-Tier suppliers (system- and model-suppliers) are located in most European supplier parks
Service offer	Typical services offered are: canteen, fire department, post office, conference rooms, medical health care, etc.
Investment-models	The three most prominent investor models are: PPP (Public Private Partnership), OEM and logistics provider are established in practice.
Operating models	The park's operator can be a subsidiary company of government authorities, the OEM or the logistics providers themselves.

Inbound-logistics	The suppliers within the park manage their inbound logistics themselves. Therefore supplier parks do not define the inbound logistics as a service offer. The transportation of goods is operated by trucks.
Outbound-logistics	The Just-in-Time and Just-in-Sequence delivery with low pre-controlled times are utilized for monorail conveyor, trolleys, automated guided vehicle systems (AGV) and trucks.

The success factor or success strategy of supplier parks lies in the synergy that results from the proximity of suppliers that are right next to the final assembly line (Sako 2005). The main focus of all activities in traditional supplier parks is the best possible design of short and integrated procurement processes without any temporary stores and additional handling costs (Sako 2005). Minimal logistic costs, uninterrupted supplies and short throughput times for OEMs are the outcome (in terms of "just in time" and "just in sequence" deliveries). The typical services in a park don't require a close cooperation among the park suppliers. Moreover, the company success does not depend on those services. Examples are as follows: canteen, fire department, post office, conference rooms, medical facilities, facilities management, etc.

4 MULTI CUSTOMER SUPPLIER PARK (MCSP)

The Multi Customer Supplier Parks or Multi OEM Supplier Park is an intelligent concept that is in the early stages of the development. Only one park in South Africa was built. The basic idea of the concept is to cut the rigid relationship between the supplier park and the OEM final assembly plant. One-to-many relations between the park and several OEMs with innovative intermodal logistic solutions are created. In order to understand the difference to the concept of a traditional supplier park the main distinctions are listed below:

Site: A Multi Customer Supplier Park is not adjacent or close to a final assembly plant with a maximum distance of a few kilometers. The site can have a distance of several hundred kilometers to the OEM plants. A requirement for the distance is the possibility of JIT / JIS delivery. Concepts of a JIT delivery, for example, are developed by DaimlerChrysler and BMW (Weyer, Spath 2001; Stautner 2001).

Parts and Suppliers: In addition to parts with high variance and/or high volume the MCSP offers a production environment for other part categories. System specialists and part/component suppliers, that have low assembly work and use complex manufacturing technologies, can benefit from a distinct advantage by being positioned in the park.

Real net output ratio: In Supplier Parks low value added activities like warehousing, sequencing, kit-building or final assembling are carried through. Due to larger quantities, based on the multi customers' scale, capital intensive machines can be cost-effectively used.

Services: In addition to the services from Supplier Parks, the MCSP concentrates on services with high cost reduction potential for suppliers, e.g. a central warehouse for all suppliers in the park, a central IT-System, operated by the logistic provider, for the distribution to all OEMs, a central C-Parts purchase and a central park marketing function – providing the chance to generate new orders form new customers.

Inbound logistics: Apart from a few exceptions Supplier Parks don't offer a consolidated inbound logistic to the clients. The consolidation with intermodal concepts (trains, trucks) opens an enormous cost reduction potential for all suppliers within the park.

Outbound-Logistics: The main objective for supplying goods to the OEMs is the principle just in time and just in sequence. However concepts like VMI (Vendor Managed Inventory) with a low stock ratio are possible. For the integrated physical connection monorail conveyor, trolleys or automated guided vehicle systems (AGV) cannot be used. Consolidated intermodal transport with trains and trucks is preferred.

Multi Customer Supplier Parks provide an environment for suppliers where they can concentrate on their core competences and obtain cost savings as well as service improvements in comparison with own sites. It implicates that suppliers have to outsource all non-core competences to the park operator. The benefits are cost-saving in logistics and services as a result of scale effects (fix cost degression, optimized capacity utilization, price degression, etc.). A demand-oriented rent of buildings will reduce high investments in a separate plant and furthermore reduce the investment risk. Moreover start-up problems like approval processes, delivery delays or problems with service business partners are minimized on account of the MCSP service offer (see table 3).

Table 3. Multi Customer Supplier Park Features.

Features	Description
Logistics: inbound, internal and outbound logistics	- Door to door logistic solution operated by the park logistic provider o Consolidated inbound transport of parts with intermodal concepts o Only one warehouse for all suppliers o One distribution system operated by the logistic provider (suppliers have no investment in IT- systems) o consolidated outbound logistics to the OEMs o Line supply services (internal material flow is operated by the park) - Logistic services like sequencing, kit-building, quality checkups, etc.
Infrastructure and services	- Turnkey solution for a fast start-up - Short product lifecycle tenancies - Central facilities management (maintenance, cleaning) - Suppliers can concentrate on their core competences - Basic park services (security, reception, canteen, conference rooms, waste management) - Advanced services (Central C-Parts purchase, personal leasing, …)
Supplier Park platform	- Market access (contact to OEMs, Suppliers) - Internal know-how transfer - Spill-over effect of best practices

A purpose-built logistics and service features for suppliers are the enormous advantages of the concept. Traditional Supplier Parks have been built for reducing OEM logistic costs and guarantee deliveries, nothing else. However the MCSP concept is tailored for the needs and requests of suppliers. The main goal is to reduce suppliers' total costs of ownership and demonstrates a convincing alternative to an own separated plant.

5 CHANCE FOR SUPPLIERS

A local production plant close to the customers could have a critical influence on the competitive advantage for suppliers. Consequently they can reduce logistic costs, i.e. transport – and stock costs. Costs per piece can be reduced substantially. Apart from the hard factors another aspect is that a local presence can have a positive effect on the perception by potential customers in the AREE (Abele; Klug; Näher 2006).

The introduced MCSP-concept takes advantage of the customer's proximity and offers an environment to companies in which they can concentrate on their core competencies while benefiting from cost reductions compared with a separate plants. Cost reductions and service quality improvements are achieved in the fields of logistics and service offers by utilizing scale effects (fixed cost degression). The concept also helps the companies to cope with the various challenges of building a new production factory. For example, the demand-orientated rent of buildings is an interesting and cost-effective alternative building a separate plant. Start-up problems referring to the deliver-ability and deliver-flexibility as well as the high efforts of building a local service network are minimized by the MCSP's service offers.

Due to its specific nature, the MCSP concept does not compete with traditional supplier parks. Furthermore it is supposed to be considered as an addition for current supplier park models. Moreover, it is a direct alternative for building a separate plant.

6 CONCLUSIONS

This article started with a description of the prospects of the automotive supplier industry in the AREE. To participate in this boom, companies can either deliver their products from their existing plants, or build a separate new factory in this region or decide to rent a factory in a supplier park. Later on the concepts of supplier parks and the innovative MCSP supplier park were explained. It was illustrated that the MCSP is an innovative alternative to a single separate plant with improved general conditions and higher cost reduction potential in the AREE. The MCSP is highly valued asset to the traditional supplier park concept.

What remains to be initiated is the building of a Multi Customer Supplier Park in the AREE to provide suppliers the best possible production environment.

7 REFERENCES

[1] Abele, E.; Klug, J.; Näher, U. (2006): *Handbuch Globale Produktion*, München Wien.

[2] Gareis, K. (2002): *Das Konzept Industriepark aus dynamischer Sicht*, Wiesbaden.

[3] Automobil-Produktion (2005): *Bindung an einen OEM aufgehoben*, in: Automobil-Produktion Mai 2005.

[4] Automotive News Europe (2006): *No trains to Trnava*, in: Automotive News Europe, July 10, 2006, P.11.

[5] Klug, F. (2003): *Erfolgsfaktor Industriepark-Logistik*, in: Zeitschrift für die gesamte Wertschöpfungskette Automobilwirtschaft (ZfAW), 2003, 3, P. 28-33.

[6] Larsson, A. (2002): *The development of Regional Significance of the Automotive Industry: Supplier Parks in Western Europe*, in: International Journal of Urban and Regional Research, 26, 2002, 12, P. 767-784.

[7] Mercer Management Consulting / Fraunhofer-Institut IML and IPA, 2004: *Fast Automotive Industry Structure (FAST) 2015 – die neue Arbeitsteilung in der Automobilindustrie*, VDA-Verlag, Frankfurt am Main.

[8] Miemczyk, J.; Howard, M.; Graves, A. (2004): *Supplier Parks in the European Automotive Industry: Agents for change. Prototyping innovative e-supply chain solutions*, in: Proceedings of the 11th EUROMA Conference 2004, P. 869-878.

[9] PSA (2006): *Company Information*.

[10] Sako, M. (2003): *Govering Supplier Parks: Implications for Firm Boundaries and Clusters*. Working paper, Oxford Said Business School 2003.

[11] Sako, M. (2005): *Who Benefits Most from Supplier Parks? Lessons from Europe and Japan faces of research*, Interview with Mari Sako, Web: ttp://web.mit.edu/ctpid/www/i13/supplier-parks.pdf from 2006-07-17.

[12] Sihn, W., Palm, D., Matyas, K., Kuhlang, P., 2006: *Automotive Region Eastern Europe - Chancen und Potenziale des „Detroit des Ostens" für Automobilzulieferer*, Fraunhofer-Projektgruppe für Produktionsmanagement und Logistik, Wien.

[13] Stautner, U. (2001): *Das Projekt Kundenorientierter Vertriebs- und Produktionsprozess (KOVP)*, in: Verein Deutscher Ingenieure (Hrsg.): Innovative Logistik in der Automobilindustrie, VDI Verlag, Düsseldorf, P. 3-8.

[14] Verband der Automobilindustrie (2003): *VDA-Empfehlung 5000, Vorschläge zur Ausgestaltung Logistischer Abläufe, Teil 2: Versorgungskonzepte*, Frankfurt a. M.

[15] Westkämper, E., Freese, J., Bischoff, J., Barthel, H., Lehnert, O., 2005: *Nutzen und Potenziale von integrierten Versorgungsstrukturen wie Lieferantenparks, Industrieparks, Versorgungszentren und ähnlichen Ausprägungsformen für die beteiligten Unternehmen*, Fraunhofer-Institut für Produktionstechnik und Automatisierung, Stuttgart.

[16] Weyer, M.; Spath, S. (2001): *Das Produktionssteuerungskonzept "Perlenkette"*, in: Zeitschrift für wirtschaftlichen Fabrikbetrieb 96/1-2, P. 17-19.

RFId Implementation in a Logistics Distribution Center

[a]Péter Németh, [b]József Marek

[a] Széchenyi István University – Department of Logistics and Forwarding H-9026 Győr, Egyetem tér 1., HUNGARY

[b] Penny Market Ltd. H-2351 Alsónémedi Po.Box: 12., HUNGARY

ABSTRACT

Motivation: The Department of Logistics and Forwarding at the Széchenyi István University in Győr, Hungary started a project in the framework Interreg IIIC REGINS (REGional standardised Interfaces for a better integration of regional SMEs in the European Economy – http://www.regins.org) with the name "Promotion of RFId (Radio Frequency Identification)". The project started in June 2005, ended in June 2006. Our partners are situated in Upper Austria, Stuttgart Region, and from Lombardy Region. Our project screens the state of the art and the ongoing development of RFId technology and processes. Special focus is on the needs of SMEs. In this paper we discuss the main differences between barcode and RFId systems, the framework of their implementation in the special case of Hungarian SME's. In the second part of the paper, we discuss our research regarding practical questions of RFId implementation at a Hungarian logistics distributions center.

Results: The main result of our research is a common view of both theoretical and practical questions regarding the implementation of RFId systems. After analysing the possible RFId trends and possibilities, we elaborated a simple framework for choosing the best RFId system to use at the given company..

Contact: nemethp@sze.hu

1 STATE OF THE ART OF RFID

Radio Frequency Identification (RFID) is an automatic identification method, relying on storing and remotely retrieving data using devices called RFID tags or transponders. An RFID tag is a small object that can be attached to or incorporated into a product, animal, or person. RFID tags contain chips and antennas to enable them to receive and respond to radio-frequency queries from an RFID transceiver. RFId tag integrated circuit is designed and manufactured using advanced and small geometry silicon processes. New advances in manufacturing process are making possible different approaches based on polymers, as a low cost alternative to the former chips. In terms of computational power, RFID tags are quite poor and contain only basic logic capable of decoding simple instructions. However, they are difficult to design, because of the challenges to manage very low power consumption, noisy RF signals or keep it operating within the strict emission regulations. When the tag enters the field generated by the antenna, it starts interacting with it and thus with the reader. The reader (mobile or fixed) emits an electromagnetic interrogation signal, which, if the tag is of passive type, charges components of transponder power supply. Following this request, tag sends to the reader its unique ID code and in case others data recorder in the memory chip.

The communication can happen in two directions (reading or writing mode) and it use radio frequency signals. When the reader receives information from RFId tag, it can temporally store them, but usually as soon as impossible all data are transmitted to the host with wire or wireless infrastructure.

2 RFID VERSUS BARCODES

Compared to barcode inventory control systems RFId technology has both advantages and disadvantages, many of which are outside of product manufacture and distribution chain applications. The main advantages are the following:

- Not requiring line of sight access to be read.
- The tag can trigger security alarm systems if removed from its correct location.
- Scanner/reader and RFId tag are not (so) orientation sensitive.
- Automatic scanning and data logging is possible without Operator intervention.
- Each tag can hold more than just a unique product code.
- Each item can be individually 'labeled'.
- Tag data can be comprehensive, unique in parts/common in parts, and is compatible with data processing.
- With the right technology a plurality of tags can be concurrently read
- It can be read only or read-write.
- There is a very high level of data integrity (character check sum encoding).
- Provides a high degree of security and product authentication – a tag is more difficult to counterfeit than a barcode.
- The supporting data infrastructure can allow data retrieval and product tracking anywhere provided the scanner/reader is close enough to the tag.
- Combined with its authentication is the ability to monitor shelf life – a societal advantage in the pharmaceutical and food industry.

- Since each tag can be unique they can act as a security feature if lost or stolen e.g. a stolen smart travel card can be cancelled.
- The technology is rugged and can be used in hostile environments such as down oil wells (heat and pressure) to carry data to remote equipment.
- The technology lends itself to being updated, for example, as a car goes through its life its service record can be electronically logged with the car.

The main disadvantages are the following:

- There is a high cost (long pay-back) for integrating RFId technology into existing inventory control systems.
- External influences such as metalwork, material dielectric properties and radio interference can constrain RFId remote reading.
- If a significant number of RFId's greater systems capabilities are implemented then the host system and infrastructure have a higher capital cost and complexity than for barcode systems.
- There are currently a range of RFId application numbering systems which need unifying to increase uptake. [The International Standards Organisation (ISO) and Electronic Product Code [EPC] Global consortium, amongst others, are working to address this issue.]
- Currently there are not internationally agreed frequencies for RFId operation (other than 13.56 MHz, which is primarily used by smart cards but can also be used by other RFId tags) and permitted scanner/reader powers differ between countries. This limits product take-up. [For example, there are significant differences between the USA and European UHF frequencies.]

Nowadays, RFID technology is still under development, standards are still converging, and costs are still being sank down in order to make the attachment of tags to individual consumer products available. However, the barcode system is deeply entrenched and will not be replaced anytime soon. This means that both barcodes and RFID systems have to exist in parallel for a long time in the future. The migration from barcodes to the RFID system will not only increase demands on system capabilities and compatibilities but also increase costs on maintenance and operation of both systems.

3 VARIOUS STANDARDS IN RFID TECHNOLOGY

Nowadays two major organizations are working to develop international standards for RFID technologies in the UHF spectrum. These two organizations are EPCglobal and International Standards Organization (ISO). EPCglobal released its EPC class 1 G2 protocol for the UHF band at the end of 2004, and the ISO released its 18000-6 in August of 2004. Both standards are still under development and are not completely compatible with each other. A unified, globally interoperable RFID standard is ideal to realize the full benefits of RFID applications. The lack of a complete and unified RFID standard has caused many companies to hesitate in adopting the RFID systems; these companies were afraid of making a commitment that might render their entire RFID system investment worthless in the future.

Regulations on radio spectrum allocation for RFID use are not unified among nations. A large portion of the UHF spectrum has already been auctioned to cellular phone service providers for high license fees by a few countries. It would be difficult, if not impossible, to buy that portion of spectrum back for RFID use. This adds complexity to the adoption of RFID for global supply chain management applications where tagged goods must often travel across borders. RFID tags which respond only to a specific UHF frequency range cannot be read in countries where different spectrum bands are allocated for RFID use. The United States and Canada can allocate the frequency band from 902 to 928MHz for UHF RFID systems because their GSM bandwidths are not located within this band.[7] Outside North America, however, the frequency band around 915MHz is almost exclusively used for wireless communications services such as GSM, PHS, GPRS or 3G. The European Telecommunications Standards Institute (ETSI) has released a 2MHz band ranging from 865.6 to 867.6MHz for Europe's UHF RFID use in July 2004. Japan has allocated a 2MHz UHF band ranging from 953 to 954MHz for RFID use in May 2005. The diversity in national spectrum allocation for RFID adds more hurdles to the growth of RFID systems in the world market. In addition to the unavailability of common spectrum bandwidths for RFID systems, power regulations and certification procedures are also incompatible from country to country.[5]

4 BENEFITS OF THE RFID TECHNOLOGY

The application of RFId in execution of logistics activities made evidence of benefit. In the following the possible results are listed for each department:

For the purchase departments:
- Better productivity
- Reduced labour costs
- Faster throughput at the receiving
- Eliminates the need of physical checks of BOL/Packaging slip
- Indicates the needs for Cross-Docking
- Case/item placement on conveyors

For storage:
- Put away accuracy & efficiency
- Less Bar Coding required
- No need to scan palettes & storage bins
- Better storage space utilisation (random location system)
- Product compatibility (e.g. Hazardous products)

For packaging:

- Picking accuracy (items and amounts)
- Productivity measurement
- Time motion measurement
- No manual scans of products

For shipping:

- More accurate shipping process
- Automatic verification through the portal of outbound dock door
- All items leaving the premises are accounted.

5 IMPLEMENTATION IN A HUNGARIAN DISTRIBUTION CENTRE

The chosen distribution centre for our survey is situated near Budapest. It is the distribution center for food and non-food products of a single supermarket-chain. This chain operates 150 supermarkets in Hungary. The storing ground is 35.000 m^2, where 1.300 different goods are stored. Every day 4.000 pallets, containing 120.000 packages, containing 3.400.000 products are leaving the centre, through 126 gates.

For the next step, we have to discuss the possible place of the tags. One tag on the pallet or on the package, or on every product. Of course the most secure solution is to place a tag on every product. It would be approx. 750 million tags pro year. If we use one tag for one package, we need approx. 26,5 million tags pro year. If tags were placed only on the pallets/transport units, 990.000 tags were needed yearly.

If we take into account, for what we need the tags, and if we see the required number of tags, the use of passive tags are obvious. On the next table we compared tag prices to the value of the product in case of all three possible levels and type of tags.

RFID-tag	Price €	Level	Value €	Share in value
active	10,00	product	0,45	2222,22%
active	10,00	package	12,57	79,57%
active	10,00	transport unit	426,00	2,35%
passive	0,10	product	0,45	22,22%
passive	0,10	package	12,57	0,80%
passive	0,10	transport unit	426,00	0,02%

From that table it is obvious, that by the current price level only the use on transport unit is economic. That means, that the goods are controlled by RFId technology between the distribution center and the shops.

We have to take into account some additional costs. As we said before, the center has 126 gates. To place a reader on every gate would cost approx. 700-800.000 €. An other possibility is to buy handheld readers. For the center 20 handheld readers were enough. The costs are 20.000 €.

The best solution for the distribution centre is to have passive RFId tags on the transport units, the reading process made with handheld readers.

6 REFERENCES

[1] Lewis, S. (2004). A basic introduction to RFId Technology and Its Use in the Supply Chain: Laran RFId White Paper.

[2] Astuti, S., & Pigni, F. (2005). A guideline to RFId application in supply chains. Report for the INTERREG IIIC Regins Resarch "REGINS RFId"

[3] Hartványi, T., Kóczy, T. L., Tóth, L. (2005). Applying intelligent methods in logistics control in 3rd International Conference on Computational Cybernetic, April 13-16, 2005, Mauritius

[4] Hartványi, T., Kovács, J. (2005). Anwendung von RFID Mitteln bei informatischen Entwicklung des Produktionsregelungssystems eines Wasserversorgungs-unternehmens in Sächsische Fachtagung Workshop „LBS und RFID – Lösungsansätze in Logistik und Verkehr", 14.-15. November 2005, Starý Smokovec, Slovakia pp. 118-124.

[5] N.C. Wu, M.A. Nystrom, T.R. Lin, H.C. Yu (2006). Challenges to global RFID adoption in Technovation (Article in Press, Corrected Proof)

[6] Accenture White Paper, (2001). Radio Frequency Identification (RFID) White Paper

[7] Finkenzeller, K., (2003). RFID Handbook – Fundamentals and Applications in *Contactless Smart Cards and Identification*, Second ed. Wiley, New York

[8] C. M. Roberts (2006). Radio frequency Identification *Computers & Security* Volume 25, Issue 1, February 2006, Pages 18-26

Organizational Learning as a Dynamic Driver for the Development of Logistics

Wolfgang Greisberger[a], Corinna Engelhardt-Nowitzki[b]

[a]Siltronic AG, Johannes Hess Strasse 24, 84479 Burghausen, GERMANY
[b]Montanuniversität Leoben, Chair of Industrial Logistics, Franz-Josef-Strasse 18, A-8700 Leoben, AUSTRIA

ABSTRACT

Motivation & Results: According to current environmental changes towards increasing complexity and volatility and at the same time highly demanding individualization and flexibility requirements from the customer side, organizational learning is getting more and more important. A company has to adequately adapt to changing requirements. This means to achieve a healthy balance between permanent and necessary change at one hand, without loosing strategic identity and credibility on the other hand due to short-sighted and inefficient oscillations. A further requirement is profitability, resulting into the need to efficiently manage and further develop explicit and implicit knowledge assets within the organization. The necessity for organizational learning is not restricted to technological or product development issues, but most notably concerns supply chain management and logistics. For a company this induces a strong need to better understand, how organizational learning has to support the development of logistics. After a short definition of "logistics" and "organizational learning" the paper will develop the prerequisites and critical success factors for organizational learning within different maturity levels of logistics, a company may be situated in. The scientific approach integrates methodological elements of individual and organizational learning, knowledge management, quality management, process optimization, supply chain management and logistics. A practical case study shows, how the theoretical results have been implemented successfully in the semiconductor industry.

Contact: wolfgang.greisberger@siltronic.com

1 INTRODUCTION

Organizational learning is getting more and more important, especially because of globalization, shorter development cycles, dynamic markets etc. Logistics are fully involved in this permanent change.

The paper provides answers to the question, what requirements a certain logistics maturity level defines for the employees and the organization of an enterprise in a "situational model of logistics". The "learning model of logistics" explains what implications are induced by transforming the challenges of each development stage into concrete requirements for organizational learning. Additionally the results of an analysis show, what kind of learning methods and didactical elements are appropriately supporting the further development of logistics according to situational matters.

The case study in the semiconductor industry explains how the classification in the "situational model of logistics" is applied, which deficits according learning were identified and how the implementation of an appropriate learning method can be driven towards actual improvement results.

2 THE SITUATIONAL MODEL OF LOGISTICS

There exists a variety of development models of logistics, e.g. [Goepfert, 01] or [SCOR, 06]. Figure 1 describes a model with four development steps, explained by their characteristics and their organization chart. The characteristics of the different steps are divided in four levels: paradigm, context, process and element.

"Paradigm" stands for shared symbols or a model used in an organization. "Context" level refers to the stakeholder, who influences the logistics. The level "processes" mean the processes in the supply chain. The content of knowledge in the logistics organization can be found in the "element" level.

The four steps of the logistics model are:

(1) Transport & Storage Logistics
(2) Coordination of process chains
(3) Coordination of value chains
(4) Flow Management

<u>ad (1)</u>: Transport & Storage Logistics is characterized by a functional organization, where logistics is not a core competence. This step is described by its goal to fulfill transfer needs [Klaus, Krieger, 2004].

<u>ad (2)</u>: By coordinating process chains, all processes in an enterprise should be balanced and optimized. Logistics in this sense is now a cross sectional function with a holistic view [Klaus, Krieger, 2004].

<u>ad (3)</u>: Value chains are supply chains "from cradle to grave" (the system gets bigger, [Bertalanffy, 03]), which are rather stable, e. g. because of long term contracts. Logistics becomes a management tool to optimize the value chain from customer point of view [Womack, Jones, 2004].

<u>ad (4)</u>: Flow Management means value chains, changing their partners as quickly as new partners can generate a benefit. The so called "Clockspeed" is fundamentally higher than in step 3 [Fine, 99]. It is a specific point of view of a supply chain, where economic coherences are explained with models of flow mechanics [Klaus, 2002].

Step		Characteristics
1: Transport & Storage Logistics 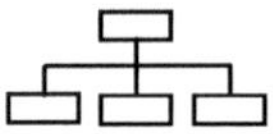	Pa	Functional thinking
	Co	React on context changes
	Pr	Workflows instead processes
	El	Working with "plan" only
2: Coordination of Process Chain 	Pa	Thinking in process chains
	Co	Reactive instead of proactive
	Pr	Enterprise wide optimization
	El	Act in process chains
3: Coordination of Value Chains 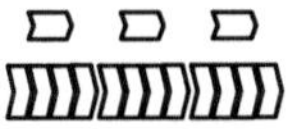	Pa	Value chain thinking
	Co	Values from cradle to grave
	Pr	Enterprise wide optimization with external trigger
	El	Act in value chains
4: Flow Management 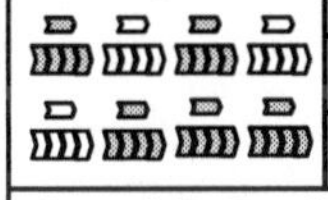	Pa	Thinking in Flow Systems
	Co	Temporary network partners
	Pr	Modular and standardized
	El	Act in value chains

Legend: Pa:Paradigm; Co:Context; Pr:Process; El:Element

Fig. 1. Situational model of logistics.

The horizon of paradigm e. g. is getting wider and wider and the development step of logistics is growing higher and higher. Organization charts show that more and more enterprises are connected to a supply chain from the beginning of a product to the final consumer, the higher the step is.

The model is called "situational" because each enterprise has to choose on its own which development step they want to become and implement respectively [Hieber, 02], [ELAAT, 04]. This decision is often driven by customer needs and requirements. Therefore the step of a company logistics is a result of interal possibilities and of customer requirements of the industry the enterprise is working in.

3 THE LEARNING MODEL OF LOGISTICS

To transform characteristics of logistics into learning requirements a learning organization has to be defined: "A learning organization is an organization skilled at creating, acquiring, and transferring knowledge, and at modifying its behavior to reflect new knowledge and insights" [Garvin, 93].

The learning requirements are again described in the four learning steps. In the coordination of process chains it is a requirement to have process learning on the process level. The case study in chapter 5 explains how to implement process learning because of problems to learn of failures in the supply chain.

(1) Learning in Transport & Storage Logistics
(2) Learning in the coordination of process chains
(3) Learning in the coordination of value chains
(4) Learning in Flow Management

The certain steps are explained with their characteristics and their organization chart.

ad (1): Learning is concentrated in the Logistics department or functional unit. Learning loops with other departments are rare. Learning loops like shown in figure 2 define the area, where a unit, an organization or a supply chain reflect new knowledge and insights. Only the main learning loops of logistics are illustrated.

ad (2): All processes of an organization are linked together in learning loops. There is rare knowledge exchange with customers, supplier or stakeholder.

ad (3): Now the value chain is connected with learning loops. Everything is orientated on customer needs and requirements.

ad (4): In this step the learning loops cover also the hole value chain. These loops have to be organized different to step 3 because value chains change their partners quicker.

Learning Step		Learning Requirements
1: Transport & Storage Logistics 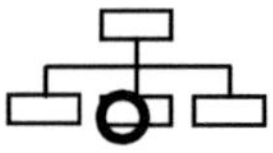	Pa	Trust organization
	Co	External triggered context learning
	Pr	Functional experience learning
	El	Passive element learning
2: Coordination of Process Chain 	Pa	Learn to think in processes
	Co	Less context learning
	Pr	Process learning
	El	Learning the elements of the enterprise processes
3: Coordination of Value Chains 	Pa	Learn to think in value chains
	Co	Learn from the Stake-Holder
	Pr	From Cradle to Grave learning
	El	Learning the elements of the value chain
4: Flow Management 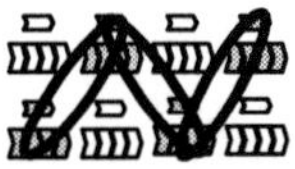	Pa	Learn to think in flow systems
	Co	Learning in changing contexts
	Pr	Process learning with changing input
	El	Learning flexibility

Legend: Pa:Paradigm; Co:Context; Pr:Process; El:Element
O :Learning loop.

Fig. 2. Learning model of logistics.

In order to be a learning organization the logistics organization also has to be skilled as mentioned to generate benefits for the company. Figure 2 shows the requirements in the levels and in circles the learning loops in the certain steps of logistics. Step-1 learning only occurs in the function logistics. By coordinating the process chain the whole organization can learn, because everybody is involved in logistics processes. Value chains learn "from cradle to grave". Flow Management makes it important to be flexible in learning, because chain partners vary.

4 METHODS OF ORGANIZATIONAL LEARNING IN LOGISTICS

Based on the definition of organizational learning, there are a lot of methods which can be used to improve learning in logistics.

Four groups of methods are evaluated in figure 3, Organizational Learning [Pawlowsky et al. 01], Knowledge Management [Davenport, 02], Quality Management [ELAKSA, 04] and Commoditization of Processes [Davenport, 05]. For the evaluation of methods see, e. g. [Klaus, Mueller-Steinfahrt, 94] or [Joebstl, 99].

The evaluation was done by using literature, discussions with experts and practical experiences.

Figure 4 shows the table of methods which can be selected. The rating "very good to use", "good to use" or "less promising" shows in which level and step a method can be used.

There are also methods which should only be used in combination with other instruments and some has to be investigated for their use in logistics.

Before applying figure 3, the organization has to evaluate its deficits or its desired improvement. There are either deficits in the step the organization is in or the organization would like to reach a higher step. In addition they have to figure out in which level they have problems.

An example for detailed results of an evaluation of a learning method is shown in figure 4. The idea behind Quick market intelligence is to install a tool with quick learning loops between receiving information, interpretating and making decisions. Regional managers of supermarkets e. g. visit their stores and the stores of the competitors from Monday to Thursday. In regular meetings on Friday they discuss new knowledge, make decisions together with the management and spread the new knowledge and decision in their stores.

	Methode	Level				Step			
		Paradigma	Kontext	Prozess	Element	1	2	3	4
Organisational Learning	Quick market intelligence	-	++	-	-	-	-	+	++
	Gamma	+	+	-	-	-	-	+	++
	Yellow pages knowledge map	-	+	+	-	-	+	+	+
	Open space technology	++	+	+	-	-	-	+	+
	Work-out	-	+	+	+	+	+	+	++
	grapeVINE	-	o	o	-	-	o	o	o
	Learning histories	+	=	+	-	+	+	+	+
	Dialoge	+	+	o	-	-	+	+	+
	Learning contracts	-	-	-	+	-	+	+	+
	Shadowing	-	-	-	o	-	o	o	o
	Learning laboratories	+	+	+	-	-	+	+	+
Knowledge Management	Communities of Practice	-	+	+	-	+	+	+	+
	Continuous Improvement Process	-	-	+	-	+	+	+	+
	Lean Six Sigma	-	=	++	-	-	+	+	+
	Failure Management	-	=	++	-	-	+	+	++
	Knowledge Balance	o	-	-	-	-	o	o	o
Quality Management	Quality Function Deployment	-	++	-	-	-	+	++	++
	FMEA	-	-	++	-	-	+	++	++
	Failure tree	-	-	-	-	-	-	-	-
	Poka Yoke	-	+	++	-	+	++	++	++
	Benchmarking	-	+	++	+	+	+	++	++
	Value analysis	-	+	+	+	-	-	++	++
	Audit	+	+	++	+	-	+	++	++
Commo-dization	Process activity and flow standards	+	++	++	+	-	+	++	++
	Process performance standards	+	++	++	+	-	+	++	++
	Process management standards	+	++	++	+	-	++	++	++

Legend: ++: very good to use; +: good to use; -: less promising; o: only in combination with other methods; =: has to be investigated

Fig. 3. Table of methods.

Evaluation of Quick market intelligence			
Level	Paradigma	-	Not useful in this level.
	Kontext	++	Very good for information sharing
	Prozess	-	Barely Process-Learning
	Element	-	Barely Element-Llearning
Stufe	1	-	Less promising
	2	-	Less promising
	3	+	Good to communicate in the value chain
	4	++	Rapid exchange of knowledge possible

Fig. 4. Quick market intelligence

5 CASE STUDY

This case study shows the three steps of improving learning in logistics by answering three questions: Which development step is the company actually in? Which deficits regarding Organizational Learning does it have? Which method should be used? The example comes from a company in the semiconductor industry. Very short development cycles and a high flexibility are characteristically for this industry.

5.1 Which development step?

The enterprise is a step 2 company as shown in figure 4 (Step 2 of situational model and learning model are shown). The ranking is based on the following facts: Thinking in or the coordination of value chains are both not existent in this enterprise. But the company is process orientated and is optimizing its internal processes.

Situational model of logistics		
Step		Characteristic
2: Coordination of process chains	Pa	Thinking in process chains
	Co	Reactive instead of proactive
	Pr	Enterprise wide optimization
	El	Act in process chains
Learning model of logistics		
Step		Requirements
2: Coordination of process chains	Pa	Learn to think in processes
	Co	Less context learning
	Pr	Process learning
	El	Learning the elements of the enterprise processes

Fig. 4. Step 2 in logistics development and learning.

5.2 Which deficits regarding Organizational Learning?

Although several knowledge-systems are in use in the company, e. g. CIP, Six Sigma, FMEA etc., information about failures did not find the way to the originator in the supply chain in proper time, e.g. from production to logistics (especially on night shift or on weekends). Furthermore the corrective action was not tracked and the failure detector did not get feedback. A failure in the following system is everything that can not be documented or solved with the existing systems.

5.3 Which method should be used?

Looking at figure 3 there are several methods to de used in step 2, like CIP (continuous improvement process), FMEA (failure mode and effective analysis) or Failure Management. The other methods that can be used in step 2 at the process level are not appropriate to deal with failures (to handle information quickly in supply chains).

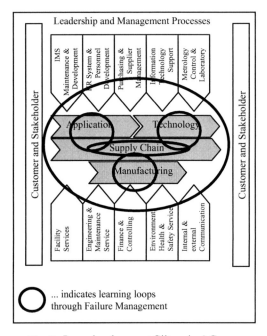

Fig. 5. Learning loops at Siltronic AG.

In a management discussion where pros and cons of the possible methods were analyzed, the solution was an Excel-VBA-Outlook based Failure Management. Everybody in the company is now able to write a Failure notification. To make sure that the notification reaches the right person, every department has an owner for the Failure

Management. This owner gets the notification and sends it to the originator.

The root cause and the corrective action have to be sent to the department responsible for the notification writer. So the time and the quality of the response can be tracked. Everybody who has a problem can have a look into the system and see how their and other problems have been solved. The time limit for response and problem solving is actually 5 working days.

In figure 5 you can see the process map of the company and the learning loops which connect different departments. In the future all departments should be connected.

5.4 Results and lessons learned

The results of the implementation of the Failure Management System are:

(1) Better communication (within and between departments).
(2) Better understanding of problems of other departments.
(3) Increase in motivation (especially on Operator level).
(4) Good documentation and reporting of failures (incl. corrective action).
(5) Output to input relation is very positive.
(6) Customers are satisfied to see this additional improvement tool in place.

Finally it is a simple tool with great benefit!

At the beginning of the implementation people were reserved, because it was again an additional tool to handle with. After using it several times it was getting common to use this tool for failures. This was caused by the attributes of the tool: simple use, fast feedback, information does not get lost in the hierarchy, clear responsibilities.

A second success factor was the positive engagement of the management. By forcing their people to use this instrument they showed their support for this tool.

6 CONCLUSIONS AND FUTURE WORK

Improving the ability to learn in logistics consists of three steps: 1. Definition of the development step of the logistics. 2. Define problems or the future step and translate it in Organizational Learning requirements. 3. Select and implement method.

It is not always obvious in which development step logistics is acting. Sometimes it can be difficult to distinguish between step 2 and 3. For this reason it is useful to discuss each level, paradigm, context, process and element, on its own and decide which method is helpful to reach the next step or solve the existing problems.

The results of the case study demonstrate how the situational model of logistics, the model of learning logistics and the evaluated methods of organizational learning can be used in practice. The case study illustrates that organizational learning can develop organizational logistics.

The actual status of organizational learning in logistics is that many organizations are situated in a step 2 or step 3 logistics. This background leaves us some questions for future examinations:

(1) Are the results of the evaluation of the methods also correct for development step 4 of logistics?
(2) Is a step 5 logistics possible in the future and what are the characteristics of this logistics and therefore what are the requirements for Organizational Learning?
(3) The focus of this work is about the manufacturing industry. What is it like in the service industry?
(4) Which learning type (Single loop learning, Double loop learning, Deutero learning; [Argyris, Schoen, 78]) is most important in logistics?

7 REFERENCES

[1] Argyris, C. / Schoen, D.: *Organizational Learning: A Theory of Action Perspective.* Addison-Wesley Publishing Company, 1978, Reading Massachusetts.

[2] Bertalanffy, v.L.: *General System Theory.* 14th edition, George Braziller, 2003, New York.

[3] T. Davenport, G. Probst: *Knowledge Management Case Book.* 2nd Edition, Publics 2002, Erlangen.

[4] T. Davenport: *The coming commoditization of processes.* In: Harvard Business Review, June 2005, 101-108.

[5] Donaldson, L.: *The Contingency Theory of Organizations.* Sage Publications, 2001, London.

[6] European Logistics Association: *Doctorate Workshop.* Deutscher Verkehrs-Verlag, 2003, Hamburg.

[7] European Logistics Association, AT Kearny: *Differentiation for Performance.* Deutscher Verkehrs-Verlag 2004, Hamburg.

[8] European Logistics Association, KSA: *Success Factor People in Distribution Centers.* Die Deutsche Bibliothek, 2004, Brüssel.

[9] Fine, Ch.: *Clockspeed.* Verlag Hoffmann und Campe, 1999, Hamburg.

[10] D. Garvin: *Building a learning organization.* In: Harvard Business Review, July-August 1993, 78-91.

[11] Goepfert, I.: *Logistics for the future*, Gabler, Wiesbaden, 2001.

[12] Hieber, R.: *Supply Chain Management – A Collaborative Performance Measurement Approach.* Vdf Hochschulverlag, 2002, Zürich.

[13] Jöbstl, O.: *Einsatz von Qualitätsinstrumenten und -methoden.* Deutscher Universitäts-Verlag, Wiesbaden.

[14] P. Klaus, U. Mueller-Steinfahrt: *The diffusion of logistics in large organisations: Playful ways towards a serious goal.* In: Proceedings of the twenty third annual transportation und logistics educator conference, 1994, 139-160, Cincinnati, Ohio.

[15] Klaus, P.: *Die dritte Bedeutung der Logistik.* Deutscher Verkehrs-Verlag, 2002, Hamburg.

[16] Klaus, P. / Krieger, W. (Hrsg.): *Gabler Lexikon Logistik.* 3. Auflage, Verlag Gabler, 2004, Wiesbaden.

[17] P. Pawlowsky, J. Forslin, R. Reinhardt: *Practices and Tools of Organizational Learning.* In: M. Dierkes, A. Berthoin Antal, J. Child, I. Nonaka: Handbook of Organizational Learning and Knowledge, Oxford University Press, 2001, New York.

[18] SCOR *Overview Version 7.0.* http://www.supply-chain.org

[19] Willke, H.: *Systemisches Wissensmanagement,* Lucius&Lucius, Stuttgart, 1998.

[20] Womack, J. / Jones, D.: *Lean Thinking.* 2. Auflage, Verlag Campus, 2004, Frankfurt a. M.

Supply Chain Strategy and Organization

Tatjana Wallner[a], Alexander Stüger[a]

[a]Upper Austria University of Applied Sciences, Campus Steyr,
Wehrgrabengasse 1-3, A-4400 Steyr, AUSTRIA

ABSTRACT

Motivation: A lot of companies are confronted with the problem how to integrate supply chain management (scm) into the strategic planning process and the organizational structure. The core advantage of scm is to align business processes with customer needs and achieve competitive advantage. The classical literature concerning strategy and structure does not consider the requirements of scm and so the full potentials cannot be achieved. Therefore the requirements have to be integrated into the strategic planning process and the design of the organizational structure.

Results: The current paper tries to close the gap between classical management and supply chain management. Based on the perceptions of classical concepts of strategic planning and organizational design it presents an approach which integrates the requirements of scm into the management functions strategy and organization.

Availability: The approach empowers companies to identify the real needs of their customers, choose the right strategic positioning, target customer segments and align the processes and the organizational design ideally according to their requirements.

Contact: tatjana.wallner@fh-steyr.at
alexander.stueger@fh-steyr.at

1 INTRODUCTION

Due to an increase of stress of competition companies are forced to locate new ways of differentiation in order to gain and sustain successful. Against this backdrop the cross company planning, managing and optimizing of processes – subsumed as supply chain management (scm) [Kuhn and Hellingrath, 2002] – is often stated as a key differentiation factor [Walther and Bund, 2001; Bruhn, 2004].

Despite the proofed relevance of modern logistics, which here is equated with scm, the real potentials are not utilized adequately yet [Cap Gemini Ernst & Young, 2002]. This is because the paradigm shift of logistics from physical goods movement to scm has not been taken into account in the classical management approaches.

In theory as well as in practice scm is often not holistically integrated into the management functions strategy, processes, organization, culture and controlling.

The Steyr Network Model (SNM), which is developed at the *Competence Center Logistics and Enterprise Networks* (Logistikum) of the *FH OÖ, Campus Steyr*, tries to close the gap between classical management and supply chain management. The SNM raises the claim to integrate the requirements of scm – customer orientation, process perspective, mutability and cross-company networking [Staberhofer and Wallner, 2006] – into all significant management functions.

The upcoming article will observe the consequences of an integration of the scm requirements into strategic planning and organizational design more deeply.

2 SCM AND STRATEGY

2.1 Classical strategic planning

A detailed screening of classical strategic planning approaches showed that although every approach is individual, the rough process of each is quite similar. Fig. 1 gives an overview about the major steps of common strategic planning approaches.

Fig. 1. Major steps of a common strategic planning approach.

By crafting the vision and mission the purpose of the company ("who we are and what we do") as well as the long-termed future goals ("where we are going") are determined [David, 2002].The environmental analysis serves to figure out risks and opportunities within a company's environment. This contains the analysis of factors of the global environment (legal, economic, ecological, social and technological factors) as well as the analysis of the company's competitive environment (for example by using the branch-structure-analysis according to Porter [2002a]).The target of the company analysis is the identification and evaluation of the company's strengths and weaknesses.

The ultimate ambition of every company is to ensure long lasting profitability [Baum et. al., 1999]. Therefore companies are forced to identify and exploit competitive advantages that differentiate themselves against their competitors [Hamel and Prahalad, 1997]. The comparison of strengths and weaknesses with risks and opportunities provides the basis to decide, the built up of which competitive advantages to focus on. The strategic positioning then depends on the target and the goals, the company wants to reach.

2.2 Consequences of the requirements of scm on the strategic planning process

(1) Customer orientation

In order to face the increasing harsh environment, customer orientation is getting more and more important [Bruhn, 2004]. But customer orientation can only become a key differentiation factor to gain competitive advantage, when the orientation on customer needs is integrated into the strategic planning process already:

- **Integration into the environmental analysis:** Additional to the analysis of the global environment and the competitive environment, there has to be special emphasis on the analysis of customer needs.
- **Integration into the company analysis:** The key performance indicator for customer orientation is whether or not the company manages to satisfy the needs of their customers better than the competitors. Therefore the capabilities of the company have to fit with the requirements of the customers [Chopra and Meindl, 2001]. As a consequence the company's capabilities have to be evaluated from the customers' point of view.
- **Integration into the strategic positioning:** According to Treacy and Wiersema not all customers have the same needs. Therefore the strategic positioning has to integrate the focusing on a customer segment with homogeneous requirements. The strategy has to be aligned to ideally satisfy the needs of the chosen customer segment [Chopra and Meindl, 2001].

(2) Process perspective

- **Integration into the company analysis:** Chopra and Meindl state that the evaluation of a company's strengths and weaknesses has to refer to the expectations of the customers [Chopra and Meindl, 2001]. Porter developed the value chain in order to identify activities that ad value to the customers and are done cheaper or better by the company then by the competitors (strengths) [Porter, 2002b]. The structured analysis of the performance of all processes that are involved in adding customer value enables the customer oriented identification of strengths and weaknesses.

(3) Mutability

The requirement for mutability can't be directly assigned to one of the steps of strategic planning. Rather affords strategic planning a high degree of flexibility and mutability in order to face the increasingly dynamic and harsh competitive environment. As a consequence, companies are becoming forced to permanently rethink and if necessary change their strategic position [Baum et. al., 1999].

(4) Cross-company networking

Companies tend to face the environmental challenges through increased cross-company integration [Walther and Bund, 2001; Bruhn, 2004]. As a consequence, the competition between single companies evolves into a competition between supply chains [Barney, 1996]. This development has to be considered in strategic planning.

- **Integration into the environmental analysis:** Porter identified the negotiation power of suppliers and customers, the threat of new competitors and substitutes and the rivalry amongst the existing companies as the five driving forces that determine the structure of a branch [Porter, 2002a]. Due to the increased cross-company networking the impact of these forces has shifted. Suppliers and customers for instance are no longer exclusively rivals but partners that target the same goals. Therefore the focus of the branch-structure-analysis has to be shifted from a company point of view to a supply chain oriented one.

- **Integration into the company analysis:** As mentioned above the focus of the company analysis is on processes that add value to customers. Until now strengths and weaknesses have been solely evaluated within company boarders. Due to cross-company networking the efficient management of the cross-company processes and the interfaces at the company boarders are becoming key factors for competitive advantage.

- **Integration into the strategic positioning:** Strategic positioning traditionally means aligning the whole company with common goals. Efficient cross-company networking therefore consequentially requires a consistent positioning of all supply chain partners.

The results of a detailed analysis of classical strategic planning approaches figured out that the above mentioned requirements are just met very generally [e.g. Treacy and Wiersema, 1997; David, 2002; Thompson and Strickland, 2002]. Especially the required alignment according to customer needs is often missing.

2.3 Implementation of customer orientation into strategic planning

The text below provides an approach, how customer orientation can be implemented in the strategic planning process (Fig. 2). The steps mentioned don't substitute traditional strategic planning but complement it.

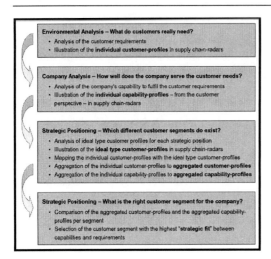

Fig. 2. Implementation of customer orientation into strategic planning.

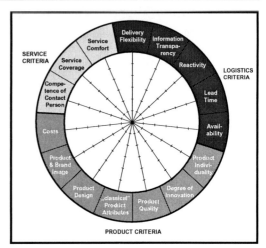

Fig. 3. Supply chain-radar.

(1) Environmental analysis – what do the customers really need?

Due to an upcoming assimilation of products in quality and function, companies are forced to generate added value for their customers. A detailed literature analysis showed that besides additional product criteria especially service- and logistics criteria are becoming key differentiation factors [e.g. Treacy and Wiersema, 1997; Christopher, 1998; Kotler and Bliemel, 2001; Chopra and Meindl, 2001]. All in all 15 core customer needs suitable for adding value were figured out.

The real importance of these needs for single customers can be figured out by classical marketing techniques like customer surveys via a structured check-list. The analysis and illustration of the so evaluated customer-profiles is executed by the developed "supply chain-radars" (Fig. 3). Because of the fact that the needs of the single customers differ quite much – while for some for example short lead times or fast reactivity are crucial, others may focus more on delivery flexibility or product individuality – the supply chain-radars have to be created individually for every single customer.

(2) Company analysis – how well does the company serve the customer needs?

The company's capability to fulfil the customer needs – measured from the point of view of the customers – is evaluated through customer surveys and analysed and illustrated via capability-profiles that are integrated into the supply chain-radars. As the customer-profiles, the capability-profiles have to be carried out individually for every single customer.

(3) Strategic positioning – which different customer segments do exist?

Due to the heterogeneity of customer needs Treacy and Wiersema state that companies can't be successful, trying to satisfy all customers. Rather they are forced to focus on a homogeneous customer segment and align the strategy on ideally satisfying the needs of the target segment [Treacy and Wiersema, 1997]. Treacy and Wiersema differentiate three positioning strategies – cost leadership, product leadership and customer intimacy. They provide a rough overview about the importance of some customer needs for each strategy but widely neglect logistics criteria. For this reason the characterisations by Treacy and Wiersema have been concretized and supplemented by logistics criteria. As a consequence of this, ideal type customer-profiles for the three strategic positions can be illustrated in the supply chain-radars. By mapping all individual customer-profiles to the most fitting ideal type-profile aggregated customer-profiles can be deduced. The aggregated capability-profiles for the segments are built the same way.

(4) Strategic positioning – what is the right customer segment for the company?

The selection of the ideal customer segment and deduced the ideal strategic position primarily depends on two factors: The profitability of the segment and the "strategic fit" between the company's capabilities (expressed in the aggregated capability-profiles) and the needs of the customers (expressed in the aggregated customer-profiles). The higher this correlation, the more appropriate is the selection of the segment as target group and accordingly the strategic positioning on satisfying these customers.

3 SCM AND ORGANIZATION

3.1 Classical organizational structure

As written above classical strategic planning approaches do not consider the requirements of supply chain management – customer orientation, process perspective, mutability and cross-company networking - enough. The same problem occurs when the organizational structure of a company is surveyed.

Even though a lot is written about new forms of organization e.g. process organization, fractal structure, network structure neither of the analyzed forms of organization can fulfil all requirements of supply chain management but they offer a lot of building blocks which help to design the structure of a supply chain oriented organization [Staberhofer and Wallner, 2006]. The reason for this is that the classical organizational theory does not share the world view of supply chain management even though particularly the process orientation has already found its way into classical organizational theory.

On the other hand a lot of research activities take place in the field of supply chain management. According to the definition from Klaus who defines the third meaning of logistics as a new 'view of the world' [Klaus, 2002], it is not purpose of this paper to discuss different methods to integrate logistics into the organizational structure of a company but to design the whole organization supply chain oriented. In this context the presented approach refers to the segmentation of the company based on logistics criteria which is discussed by many authors [for this purpose look at Klaas, 2002]. This theory place emphasis on the thought that a company achieves success when the structure is subdivided into parts with the same logistics criteria. Payne and Peters who have developed an approach for logistical segmentation largely based around product characteristics argue that the building of organizational segments on the basis of logistics criteria has got positive impact on the prosperity of the company. [Payne and Peters, 2004]. But the logistical segmentation does not deal with the integration of the built segments in the formal organizational structure enough.

On this account the approach presented in this paper combines the research activities in the field of classical organizational theory and supply chain management. Furthermore it shows how the described steps to develop a supply chain strategy can be transferred into the organizational design of the company so that strategy and structure are in line.

3.2 Implementation of supply chain requirements into the organizational design

The following passage ties in with the description of the implementation of customer orientation into strategic planning.

(1) Building specific supply chains – which supply chains are crucial for the organizational design?

In the literature you can find different models for a segmentation based on logistics criteria. The characteristics and the borderline of the built supply chains are not equal but there is unity in the goal to design process oriented value added segments which have got different requirements.

The criteria (service-, product- or process-criteria) on which the segmentation is based differ [Klaas, 2002].

In this paper the segmentation is based on the customer requirements described in the supply chain strategy expanded about market criteria and core competences. The strategic positioning itself only determines which customer needs the company wants to fulfil. It says which needs are very important for the satisfaction of the customer. This has got great influence on the organizational design e.g. if the company decides for the strategy of product leadership criteria like product individuality or degree of innovation has to be fulfilled by the company.

For the supply chain oriented design of the organization specific segments called supply chains in this paper have to be built. *A supply chain is defined as a special set of processes which fulfil special customer needs.*

Therefore the requirements of the customers which are visualized in the "supply chain-radars" have to be concretized (e.g. lead time of two days) and it has to be analyzed how the needs influence the processes. As supply chain management comprehends the product development, order acquisition, product planning, procurement, distribution and recycling processes [Thaler 2001] these processes has to be considered. The results of this step are defined supply chains. The defined supply chains are the basis for the organizational design of the company.

(2) Definition of the organizational design: what is the right supply chain oriented organization?

- The so built supply chains are formalised in organizational units on the highest level of organizational design. Because of the fact that the units are built based on different customer needs the requirement customer orientation is fulfilled.
- The supply chain oriented organization does not follow the classical organizational perception which focuses on the structural organization and disregard the process perspective. In this approach the process-flow is the basis for the organizational design. Thus the supply chain oriented organization is segmented primarily by processes to achieve a high degree of process orientation on the next level of organization. As already mentioned the built supply chains have got different requirements and therefore a different structure of the processes. But it can occur that two ore more supply chains have got one ore more processes in common. The method developed by Schulte-Zur-hausen [2002] is an adequate tool to find this out. First the processes get defined for the on customer requirements based supply chains. Then the processes of the different supply chains get compared and if processes are equal in one ore more supply chains for synergetic reasons it is better when these processes do not get assigned to one supply chain but build an own organizational unit (e.g. when different supply chains have got the same procurement process an own procurement unit should be built).
- The supply chain oriented organization also has to be very mutable so that the company can react flexible in case of market changes. To react on high market dynamics a high degree of decentralization of the organizational units is appropriate [Hill, 1994]. Especially for the process based organizational units, which are assigned to one supply chain, decentralization is suggested. The organizational units based on processes which are the same in one ore more supply chains should be centralized.
- Another requirement of supply chain management is cross-company networking. Especially the decentralization of organizational units and the segmentation of the organization based on standardized processes allow an easy integration of different companies. So the above described suggestions for developing a supply chain oriented organization support companies to become network-compatible. This is the prerequisite

for supply chain management which does not end in the own company.

The steps towards a supply chain oriented organization are combined in figure 4.

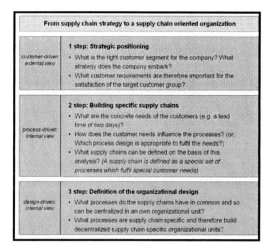

Fig. 4. From supply chain strategy to a supply chain oriented structure.

The design recommendations for a supply chain oriented organization starting from the definition of a supply chain strategy are a first step towards successful supply chain management. This approach tries to develop effective supply chain management from the view of one single company which regards the needs of the customers and the abilities of the suppliers. To tap the full potential of supply chain management future research work has to focus on a model to conjointly develop a network strategy and network organization among all supply chain partners. The research activities of the *Competence Center Logistics and Enterprise Networks* (Logistikum) of the *FH OÖ, Campus Steyr*, concentrate on the development of such a model with the Steyr Network Model (SNM).

4 CONCLUSION

Achieving strategic fit between the needs of a homogeneous customer segment and the company's capabilities to fulfil these needs is a central point of the integration of scm into strategic planning. The ideal positioning on a specific customer segment depends on the attractiveness and the level of strategic fit.

The needs of the target customers are the central criteria for a further segmentation on top level of organizational design. On the lower levels of organizational design, processes are focused. Moreover the structure of a supply chain oriented organization hast to assure flexibility in case of environmental changes (mutability) and the capability to support cross-company integration. Decentralization and network ability are important features for successfully doing so.

5 REFERENCES

[1] Barney, R. (1996) *Business logistics management – planning, organizing and controlling the supply chain*, Upper Saddle River.

[2] Baum, H.-G. et. al. (1999) *Strategisches Controlling (2. Aufl.)*, Stuttgart.

[3] Bruhn, M. (2004) *Das Konzept der kundenorientierten Unternehmensführung*, in: Hinterhuber, H. and Matzler, K. *Kundenorientierte Unternehmensführung (4. Aufl.)*, Wiesbaden, S. 33-65.

[4] Cap Gemini Ernst & Young (2002) *Networked Value Chain Studie – der Aufbau adaptiver Wertschöpfungsnetzwerke: Bedeutung, Nutzen und Erfolgsfaktoren*, Berlin.

[5] Christopher, M. (1998) *Logistics and supply chain management – strategies for reducing cost and improving service*, London.

[6] Chopra, S. and Meindl, P. (2001) *Supply chain management – strategy, planning and operation*, Upper Saddle River.

[7] David, F. (2002) *Strategic management - concepts and cases*, Upper Saddle River.

[8] Hamel, J. and Prahalad, C. (1997) *Wettlauf um die Zukunft*, Wien.

[9] Hill, W. et al (1994) *Organisationslehre 1 (5. Aufl.)*, Bern u. a..

[10] Klaas, T. (2002) *Logistik-Organisation – Ein konfigurationstheoretischer Ansatz zur logistikorientierten Organisationsgestaltung*, Wiesbaden.

[11] Klaus, P.(2002) *Die dritte Bedeutung der Logistik*, Hamburg.

[12] Kotler, P. and Bliemel, F. (2001) *Marketing-Management – Analyse, Planung, Umsetzung und Steuerung (10. Aufl.)*, Stuttgart.

[13] Kuhn, A. and Hellingrath, B. (2002) *Supply Chain Management – optimierte Zusammenarbeit in der Wertschöpfungskette*, Berlin, u.a.

[14] Payne, T. and Peters, M. (2004) *What Is the Right Supply Chain For Your Products?*, in: The International Journal of Logistics Management, Volume 15, Nr. 2, 77- 91

[15] Porter, M. (2002a) *Wettbewerbsstrategie*, Frankfurt.

[16] Porter, M. (2002b) *Wettbewerbsvorteile*, Frankfurt.

[17] Schulte-Zurhausen, M. (2002) *Organisation (3. Aufl.)*, München.

[18] Staberhofer, F. and Wallner, T. (2006) *Supply Chain-orientierte Unternehmensorganisation*, in: Supply Chain Management, **2**, 47-52.

[19] Thaler, K. (2001) *Supply Chain Management (3. Aufl.)*, Köln.

[20] Thompson, A. and Strickland, A. (2002) *Strategic management – concepts and cases (13. Aufl.)*, Chicago.

[21] Treacy, M. and Wiersema, F. (1997) *Marktführerschaft – Wege zur Spitze*, Frankfurt u.a..

[22] Walter, J and Bund, M. (Hrsg.) (2001) *Supply Chain Management – neue Instrumente zur kundenorientierten Gestaltung integrierter Lieferketten*, Frankfurt am Main.

The research results in the field of supply chain oriented organization were developed in the FH*plus* project ProOrga.

Critical Success Factors for Supply Chain Collaboration and Supply Chain Governance in Complex and Dynamic Environments:
A "Design for Agility"

Corinna Engelhardt-Nowitzki [a], Helmut E. Zsifkovits [a], Elisabeth Lackner [a]

[a] Montanuniversität Leoben, Chair of Industrial Logistics,
Franz-Josef-Str. 18, A-8700 Leoben, Austria

ABSTRACT

Motivation: Successful supply chain management (SCM) does not only mean excellent order and service fulfillment capability within complex and volatile environments. There is a rising need to efficiently and flexibly steer open sections of the supply chain (SC) that are undergoing permanent parameter and configuration changes. As a consequence of prior or actual outsourcing activities the supply network is often highly fragmented and involves a rising number of (legally independent) partners. According to relevant economic approaches (e.g. transaction-cost economy, property-rights theory, principal-agent approach), a working relationship needs an ex-ante contracting and also ex-post steering activities ("governance"). In addition to formal agreements, a number of informal forces are influencing the stability and efficiency of collaboration.

A company therefore has to know the critical success factors of SC collaboration and SC governance and to subsequently operationalize generic considerations according to specific situational requirements.

The main research question of this paper is: What are the critical success factors of SC configuration and SC governance within a dynamic SC network? What concrete SC design recommendations can be derived regarding process structure, process performance and process steering principles? A special focus lies on the question how to "design a supply chain for agility".

Results: Basic insights from economic theories, SCM, process management, logistics and risk management will be combined and transferred into practically applicable concepts. The final paper will provide critical success factors that have been derived from an interdisciplinary point of view. The scientific approach is combining findings from disciplines that are closely related to SCM with analogies from other disciplines (e.g. biology). In a first step ("micro-level") several types of SC collaboration constellations will be analyzed for their characteristics, prerequisites and adequate steering principles. Secondly the synthesis of the single constellations ("macro-level") will find answers to the aforementioned research questions. A "design for agility" implies two perspectives: One is to achieve an agile, but still strategically well-considered SC (re-)design process. The second is to (re-)design agile – or in other words – fast, responsive and adaptive, but still stable and productive process chains.

Availability: A company retrieves concrete recommendations for SC (re-)design based on a standardized process reference model. Commonly used process reference models will be extended within the paper especially regarding a "design for agility" in order to better address relevant SC phenomena such as the bull-whip effect or exaggerating demand customization claims.

Contact: corinna.engelhardt@mu-leoben.at

1 TRENDS IN SUPPLY CHAIN MANAGEMENT

In order to survive an organization must meet two requirements: It must show an acceptable economic performance and it must continuously adapt as necessary to changing environmental conditions. Agility is the competency of an enterprise to deal with changes in the critical competitive practices of its business sector. Agility is herewith relying on concepts like asset and cycle time reduction, mass customization, virtual enterprise, reengineering, learning organization, and lean production (Dove, 1995). The two concepts discussed in order to fulfill the requirements defined above, agility and leanness, can be seen as a dichotomy (Engelhardt-Nowitzki and Zsifkovits, 2006b). A lean system faces no waste of time or material. Capacity and resource utilization are high, but there is no opportunity for change. An ideal lean system is efficient and static. An agile system can adapt to external signals. Its utilization is below 100%, providing the necessary resource buffers for adaptation purposes. It is an effective, dynamic system.

We extend the apparently dichotomic concepts of agility and leanness from a single enterprise perspective to a SC perspective. Global supply chains are part of complex and volatile environments undergoing permanent parameter and configuration changes. The SC network is typically highly fragmented, involves high risk and a number of partners that can be legally independent. SC collaboration forms that aim at adaptiveness and stability at the same time has become a critical requirement for survival. "Design for Agility" in this context has two goals (Engelhardt-Nowitzki and Zsifkovits, 2006b; Haberfellner and de Weck, 2005):

- Agile SC design, meaning to design or adapt the process chain architecture with a very flexible design approach.
- Design of an agile SC, that finds a balance between leanness and agility.

The second aspect leads to a new paradigm: The overall objective may not anymore be a short-term efficiency approach but has to be turned towards long-term effectiveness. This changes basic enterprise-internal functions such as target agreements, resource allocation principles, risk management and also influences the way of interorganizational (supply chain) collaboration.

2 A DEFINITION FOR SUPPLY CHAIN COLLABORATION

SC collaboration is defined as a means by which all companies involved in the SC are actively working together towards common objects. It is characterized by sharing information, knowledge, risk and profits (Mentzer et al., 2000). Horvath (2001) as well as Simatupang and Sridharan (2002) see the motivation in SC collaboration in the improvement of overall SC performance while Mentzer et al. (2000) see the benefits in inventory reduction, improved customer service, more efficient human resources, reduced cycle times as well as considerable financial benefits but also certain non-financial benefits.

In terms of "governance" which in this article stands for the procedures and systems by which an organization or society operates with its SC partners, collaboration can be interpreted as the medium to encourage all parties involved in the SC to actively work together on an (at least partially) common goal. Sauvee (2002) defines governance further as "an institutional structure for which the role is simultaneously to define a process of adjusting durably a collective action (or strategy) between autonomous entities through the establishment of a 'private order' (Williamson, 1996[b])" and to design mechanisms (either contractual or non-contractual) enabling the assurance, at the lowest cost, that the individual behavior of partners follows the rules for collective action". This definition on corporate level shows the relationship amongst the stakeholders in a certain environment. It is a set of processes, customs, policies and rules that directly affect the involved SC parties. Another aspect of governance is the so called global governance, defined as the regulation of interdependent relations in the absence of an overarching political authority (Rosenau et al., 1997) as well as collective efforts to identify, understand or address worldwide problems that go beyond the solving capacity of individual states (Weiss, 2000). Global governance is not discussed in this paper.

3 LEVELS IN SUPPLY CHAIN COLLABORATION

SC collaboration distinguishes different levels of scope. Depending on these levels, influences and controllability differ widely. Pfohl (2003) uses a common distinction: Macro-logistic systems apply a macro-economic perspective, e.g. the national system of goods transport. Micro-logistics looks at the logistics systems or sub-systems of individual enterprises. This level focuses on intra-organizational systems. Meta-logistics in Pfohl's sense is a level between macro- and micro-logistics. The scope is on logistic sub-systems, e.g. the cooperation between suppliers, manufacturers, retailers and logistics providers in a selected value chain. This perspective is inter-organizational. Even if Pfohl does not use the term supply chain, this describes parts of or a supply chain as a whole.

Christopher and Peck (2004) uses a similar approach for classifying SC risks. Adapted from a model that was originally developed by Mason-Jones and Towill (1998), he identifies three perspectives, which are further divided into five categories. The following table compares both approaches:

Pfohl	Christopher
Micro-level	Internal to the firm – Process – Control
Meta-level	External to the firm but internal to the supply chain network – Demand – Supply
Macro-level	External to the network – Environmental

Fig. 1. Levels of scope.

This paper will answer the following questions:
- What are the factors that are decisive in successful SC collaboration and governance (critical success factors)?
- What different role do these factors play at micro-, meta- and macro-level?
- What are promising solution approaches and supply chain design recommendations and which analogies from other disciplines can be drawn additionally?

4 BASIC PRINCIPLES OF COLLABORATION

Collaboration refers to a process where people work jointly together based on a contract that defines
- achievement and reward,
- rights and duties of both contracting parties, and
- execution modalities.

This applies to people as well as larger groups and collectives/societies. In the micro-level, those contracts are only between natural persons and humans as part of an organization (the individuals in a contract) whereas on a meta-level the contracts are between organizations and teams. This can be organizational forms like joint ventures, private and public limited companies, franchising or licensing organizations etc. The macro-level would then include contracts between international organizations as well as countries and communities (e.g. EU, NAFTA, ASEAN, …). "Contracts that direct decisions toward the interests of residual claimants also add to the survival value of organizations" (Fama and Jensen, 1983). As already Coase (1937) says, it is not possible to write complete ex ante contracts. This emphasizes the importance to not only adequately allocate the rights but also power and control in order to enable reasonable operations during contract duration.

4.1 Property Rights Allocation

Cooter and Ulen (1988) define property rights as a comprehensive list of what people may or not may do with their own resources. The owner is free to exercise his rights over his property, but is not obliged thereof by law. The legal conception of property is, then, that of a bundle of rights over resources that the owner is free to exercise and whose exercise is protected from interference by others. Property creates a zone of privacy in which owners can exercise their will over things without being answerable to others.

Ownership is associated to the possession of a wide assortment of use rights that range from the traditionally recognized rights of usus, fructus and abusus first identified by Roman law (Nicita et al., 2005). The term usus refers to the right to use property, the term fructus indicates the right to receive income from property, and the term abusus identifies the right to dispose of property by transforming it, transferring it, or even destroying it (Furubotn and Richter, 1997).

Property rights refer to the sanctioned behavioral relations among economic agents. Therefore they use valuable resources. The rights range from defining access and use of natural resources to defining the nature of market exchange and to work relationships within firms. No matter of the nature of the allocation, property rights must be clearly specified and enforced to be effective, and the degree of the specificity depends upon the value of the asset covered (Libecap, 1999). "... property rights are not absolute and can be changed by individuals' actions ..." (Barzel, 1997, p. 4) and therefore the correlation between the specific partners in the SC and the associated interactions between them are decisive. Property rights can be referred to as natural persons in the micro-level, organizations of different types in the meta-level and international organizations and countries on a macrobasis. The differences between those levels can be seen in the complexity of the property rights regarding the relevant areas of application and in the intersections between relationships and contracts on all levels.

A modern "supply chain design for agility" has to pay respect to the property right allocation as follows: Involved (SC) parties will seek for their optimum, possibly in an opportunistic way (Williamson, 1996a). SC collaboration contracts and governance mechanisms therefore need to motivate the partners to react quickly and flexibly on one hand and therefore to establish a decentral decision ability. Applicable approaches how to build strategic networks within this conditions are shown in detail in Jarillo (1988 and 1993).

4.2 Principal Agent Considerations

Per economic definition the principal agent theory refers to a variety of ways in which agents, are linked by contractual arrangements to a firm. These can include organizational and capital structure, remuneration policies, accounting techniques and risk-taking attitudes (Jensen and Meckling, 1976; Watts and Zimmermann, 1986). "An agent is anything that can be viewed as perceiving its environment through sensors and acting upon that environment through effectors." (Russell and Norvig, 1995). Examples for principal agent relationships are:

- Capital seeker and investor (macro)
- Subordinate and supervisor (micro)
- Proprietor and manager (meta)

In a supply chain context often a so-called "focal enterprise" is dominating the chain, herewith taking the role of a principal. Asymmetric information and different preferences may lead to distrust and hazard (Kaluza et al., 2003).

Milgrom and Roberts (1992) identify four basic principles of contract design:

- the Informativeness Principle,
- the Incentive-Intensity Principle,
- the Monitoring Intensity Principle and
- the Equal Compensation Principle.

The Informativeness principle states that any measure of performance that reveals information about the effort level chosen by the agent should be included in the compensation contract.

The Incentive-Intensity principle states that the optimal intensity of incentives depends on four factors: The incremental profits created by additional effort, the precision by which the desired activities are assessed, the agent's risk tolerance and the agent's responsiveness to incentives.

The Monitoring Intensity principle is similar to the above principle and in this case employees choose from monitoring intensities.

The Equal Compensation principle states that activities equally valued by the employer should be equally valuable (compensational terms) to the employee.

In literature a lot of ex ante and ex post solutions for principal agent problems in the SC context are discussed with emphasis on the persuasibility of the results from the agents view and on the observability of the agent from the principals view (refer e.g. to Kaluza et al., 2003, who modify a model proposed by Picot and Böhme, 1999). Most proposals try to lower information gaps, herewith taking into account agency costs.

SC collaboration design and governance have to pay respect to the mentioned influence factors in order to avoid residual loss from the view of the principal (Jensen and Mecklin, 1976). The requirement of agility in volatile environments definitely excludes the (assuming a complete contract at least theoretically given) possibility of a complete monitoring. The above given recommendation for a decentrally oriented property rights allocation enforces this problem further. Alternative mechanisms, that could (isolated or combined) substitute full control besides market- or hierarchy-contracts, that are mainly based on authoritarian influence, could e.g. be:

- trust or relationship, e.g. in a clan-organization (Ouchi, 1980) or in strategic networks (Jarillo, 1988 and 1993)
- reputation or legitimacy, e.g. a brand name
- the offer of a unique asset that can not be substituted easily, e.g. a technical ability or a granted patent
- common interests ("win-win-constellation"), e.g. subcontracting in highly variable markets, where the principal flexibly equalizes utilization variations through subcontractors, whereas the subcontractor balances demand fluctuations from several principal contractors with each other (Eccles, 1981)
- incentives or penalties, if conditions are traceable with reasonable effort

In this paper we especially propose hybrid arrangements (market – hierarchy), such as alliances, franchising models, collective trademarks etc. in order to combine decentral agility with collective advantages, synergies and strengths.

Besides a possible hazard, typically the risk affinity of principal and agent is different (Sappington, 1991). The more self-organizing, decentrally organized and therefore adaptive a supply chain is designed, the more risk decisions will be taken by an agent. Risk affinity differences could possibly be adjusted by installing appropriate incentive mechanisms.

4.3 The Transaction Cost Approach

Robert Coase (1937) was the first to introduce the concept of transaction costs: "The main reason why it is profitable to establish a firm would seem to be that there is a cost of using the price mechanism." Milgrom and Roberts (1992) derive the characteristics of transaction costs from motivation and coordination costs. Where motivation costs reflect the incentive structure within transactions, therefore the activities have to run balanced when the coordination costs are derived.

Transaction costs do not only derive in correlation with completed but also with not accomplished transactions. The extend of transaction costs depends on the following attributes: (Migrom and Roberts (1992) p. 30-33)

- Asset specificity
- Frequency and duration
- Uncertainty and complexity
- Difficulty of performance measurement
- Connectedness to other transactions

The common objective of minimum transaction costs definitely supports the lean concept, but may be misleading, if supply chain operations require a high extend of adaptiveness.

Recapitulating the three economic approaches (4.1-4.3) have led to the following conclusions regarding a "design for agility" in the context of SC collaboration:

- property rights have to be allocated in a way that motivates the related parties to

- react quickly and flexible based on a decentral liability assignment
- the adequate agreement of power and control has to be included in contracts and governance conventions
- if full information and control remains unrealistic due to high volatility, alternative liaison forces have to be established in the business model
- aspects that are related with increasingly decentral and self-organizing organizational forms, such as diverse risk preferences have to be compensated by appropriate collaboration regulations
- the strict enforcement of the (lean) transaction cost criterion may cause less optimal constellations for the requirement of agility.

The complexity enormously rises from mirco- to macro-level. From the view of a company a valid complexity reduction is to concentrate on the mirco- and meta-level. This means to internally define adequate processes, rules and standards that avoid inefficiency and waste (lean) but provide resource and eventually knowledge buffers (agile). On the meta-level the most critical factor is a benefiting business constellation according to the arguments we have discussed above. The concrete SC-design aspects to be discussed in detail will be shown below.

5 ANALOGIES FROM NATURE – A SHORT EXCURSUS

One can learn a lot about dynamic flow-systems from nature (at length compare Engelhardt, 2001). Concentrating on the ability to adapt to changes but still having to be efficient, for example the biological principle of evolution promises useful hints: Living organisms have to provide good and efficient survival mechanisms. For the example of the mammals this works as follows:

Long-term adaptation and development of a species follows the laws of evolution (see e.g. Dieckmann et al., 2004). Individuals experience finite life-cycles within a common gene pool that keeps the species stable regarding successful attributes through the principle of heredity. Adaptability is provided through gene mutation and selection and environment-caused modifications. The recombination of recessive attributes enables further changes. A certain extend of redundancy has to be accepted in order to provide an appropriate variability. Critical survival factor is an adequate variation rate.

Medium-term deployment is organized by the hormone system (see e.g. Norman and Litwack, 1997). The effect of centrally and de-centrally produced substances is specific according to situational requirements. Underlying mechanisms are recursive enzymatic control, feedback-loops and self-similarity. The progression can be cyclical or can pass through structural changes possibly including sudden growth spurts and regressive development stages.

The ability for a short-term capacity of reaction is realized by the nervous system. A high responsiveness is enabled through trigger-induced activities and reflex arc processing (see e.g. Mumenthaler and Mattle, 2006). Based on a widely ramified and fast information network that has got short routes and uses a simple, universal code-structure. Some processes are controlled consciously. Others are executed unconsciously and don't involve a central control entity. Main working principle is a set of predefined elicitors (conditional and unconditional) that can be addressed either externally or internally depending on a predefined threshold.

What are the conclusions from this analogy? Long-term SC development has to concentrate on the strategic network and chain configuration (Klaus, 1994). Central issues are, what market(s) should be involved according to market dynamics and the design of the product and service assortment. A specific SC topic is, what type of and which specific partners should be integrated, how the relationship(s) should be set up legally and how the operative processes can be coordinated. According to the mentioned evolution principles an adequate adaptation rate has to be reached. Change is provided through finiteness, which gives prove to the SC paradigm-shift from continuance to variance. The analogy gives prove to the fact that leanness can only be driven towards a critical extend: All areas that require agility have to take into account a certain amount of redundancy and a lower resource utilization to stay maneuverable. Not limited to coincidental modifications, progressive and regressive phases can be accomplished systematically exposing the relevant organizational units to a competitive environment that targets at long-term survivability, not short-term economization.

Medium-term supply chain alignment aims at process flow programming and rationalization (Klaus, 1994). According to the principles of the hormone system, the SC design should utilize closed-loop mechanisms, self-similar structures, and recursive steering mechanisms in order to achieve a low management effort.

Short-term supply chain responsiveness with a still acceptable process efficiency can be reached, if the prerequisites of an agile process mobilization and regulation (Klaus, 1994) are provided: Operative tasks should be assigned to competent de-central units, having a direct link to a central steering entity, the "distributing center". The information flow between local and central units should be based on the principles of reflex circuit processing and use an enterprise-wide common code. Early warning and appropriate escalation procedures have a high priority.

Further examples where and how SC design can learn from nature can be identified easily, although a detailed analysis goes far beyond the scope of the paper on hand. More details and examples can be found in Engelhardt (2001), Ferger (1997) or other research works that explore analogies between logistics and nature.

6 CRITICAL SUCCESS FACTORS

In any organization or organizational unit certain factors can be identified that are essential for the success of that organization or unit. If these objectives are not met, achievements can not be guaranteed and the organization may fail (Huotari and Wilson, 2001). According to Rockart (1979) critical success factors (CSF) are defined as "... the limited number of areas in which results, if they are satisfactory, will ensure successful competitive performance for the organization." CSF can be seen as (Spath et al., 2006, p. 64):

- abilities a company has to provide
- available human and material resources
- issues that have to be completed
- competitive rules that have to be analyzed.

Looking at the so far discussed SC collaboration and governance aspects from a more operative perspective, three critical factors can be assumed that have a close relation to process performance and agility:

- complexity issues, as growing complexity often decreases adaptiveness (Engelhardt-Nowitzki and Zsifkovits, 2006a)
- the technological base that enables the material and information flows at all
- communication and trust aspects among involved human parties.

Poirer and Quinn (2003) define a list of the ten most important SC success factors, that includes the item "supply chain collaboration". Among other more generic issues (e.g. management commitment or project management ability) this list further mentions the operability of technology enablers and trust between the involved people, herewith proving the above assumed factors as critical. All three identified critical success factors – technology, trust and complexity – will be analyzed further below.

6.1 Technology related CSFs

IT is the driving unit in a well functioning and flexible supply chain. The main CSF's in the field of IT are the development of powerful hardware (e.g. RFID), the integration of common software tools and particularly applications (e.g. ERP, BWH) to steer the material flow and international applicable uniform standards. Key indicators therefore are the number of system interfaces and equipment compatibility. Furthermore it is important to assure an interoperability in terms of standards (GS1, EAN, automotive standards, ...) as well as interchange formats for all members in a functional supply chain, to assure a well documented communication basis. It must be possible for all members in the supply chain to store and retrieve relevant information. Relevant key indicators for this case suggested by the authors are: Operator friendliness, defect frequency, number of queries due to misunderstandings, detection of bottlenecks, and consolidation of conflicting technology .

Data models and algorithms with shared data and information in certain applications (e.g. EDI, VMI, ...) are further CSF's directly linked to applicable uniform standards. The best collaboration concept, contract or governance construct can not perform well, if technological shortcomings are permanently causing bottlenecks, failures and redundancies within the information or material flow.

In some industry sectors additional technology-components (transportation, handling of goods, ...) play a major role. If this is the case, those may also gain the role of a CSF according to situational requirements and conditions.

While technology can be seen as a rather influenceable factor in the supply chain, the behavioral aspects of collaboration between different partners and organizations are not as simple to modify as the infrastructure issues: Aspects like communication and trust are more difficult to manage within a dynamic supply net system.

6.2 People related CSFs

Collaboration also concerns the soft factors within a supply chain. This means, that is has to be dealt with differently than other, hard aspects. People and their reactions can never be programmed or planned, but only lead into a certain direction. A key success factor for collaboration in a supply chain is the equal balancing of the interests of different partners. Therefore the choices of the right partners and the collaboration layout have to be put down upfront. For those aspects in the supply chain the following key indicators can be identified: Clear definition of roles and policy outlines – especially collaboration decisions and collaboration conditions.

Another important aspect in the field of collaboration is the communication within units and organizations. Various influences can change the communication style, the communication rhythm and so on. It has to be identified if there are different cultures, languages or levels / directions of education involved, that could lead to misunderstandings which could further cause errors or even failures.

The organizational structure of a unit or organization plays an important role. This concerns the formal organization as well as the informal factors such as mutual trust and respect between the partners (this is even more important if there are different companies involved). A further issue is to provide collaboration partners with the appropriate and complete information. Mistrust or security concerns could again cause unwanted process failures. The overall information basis can be influenced dramatically, in consequence there will be no further chance of coexistence and equal partnership. Those factors are technology-wise indicated by common shared information data bases, network settings and performance measurement systems. Again it can be stated that infrastructure problems can be a bottleneck, but can be handled much easier than social matters.

For partners within a certain network it is important that there are clear defined and communicated competences between the partners. One unit/person always has to be responsible for certain topics. Those decisions need to be made traceable, which in an agile environment asks for a certain flexibility and adaptability of all members in the collaboration. It is not always helpful to use monetary reward systems for e.g. managers, since they are just the peak of the iceberg and only to a certain degree responsible for the implementation of the factor (Deming in Hoogervorst et al.).

In the near past, there have been a few other rewards systems been tested towards applicability (e.g. team events) which "have a positive impact on levels of participation and lead to organisational or task effectiveness and performance" (Raduescu and Heales, 2004, p.18).

Collaboration should also not be seen as an end in itself but should maximize the customer benefit (including co-workers, distributors, share- and stockholders, investors, politicians and also media).

6.3 Complexity related CSFs

Complexity is a reality in supply chains, resulting from a high variety of products, components and associated processes, the disparity of customer requirements, irregular patterns of ordering, and the uncertainty of supplies. Engelhardt-Nowitzki and Zsifkovits (2006a) propose a systematical approach how to analyze SC complexity. Among other issues they introduce the aspect of "consecutive complexity", meaning that even if an aspect is not a complexity driver itself, it nevertheless may increase complexity elsewhere in the value chain. This effect could be quite relevant in the context of collaboration and governance: At one side a contract or a collaboration procedure may be complex itself. At the other hand even a simple agreement in a contract may cause a high extend of consecutive complexity within the related processes. A CSF analysis must therefore not be restricted to the assumed complexity driver but also look at consecutive effects.

The following approaches can be used in order to control complexity:
- Critical Business Practices Framework: Identify areas that lack sufficient attention, yet are critical in today's competitive environments
- Reduce product variety and customer requirements disparity through standardization and modular design
- Increase component commonality
- Increase process similarity, e.g. by applying process reference models for supply chain design and configuration
- Apply design principles from biology and nature that have proven their applicability in a context that requires agility and leanness
- Increase order stability through delivery models (e.g. VMI)
- Reduce supply uncertainty through better forecasts.

As one further aspect Bliss (2000) mentions a target conflict between the static potential of an enterprise that increases agility but at the same time rises complexity; for the dynamic complexity he finds an opposite coherency: the less alternatives, the less adaptation time is necessary for adaptation, which means a better survivability. The consequence for the analysis of complexity related CSFs is, that one has always to achieve a holistic overall picture before corrective actions can be defined.

7 CONCLUSIONS

Recapitulating it can be said that the critical success factors regarding SC collaboration and governance in environments that require agility are quite manifold. The differentiation of three levels (micro-, meta- and macro-level) reduces complexity. Main approaches that help to analyze supply chain constellations regarding collaboration are the property rights theory, the principal agent approach and the transaction cost theory. The definition of the relevant business model(s) within a supply chain is mainly determined by considerations that can be derived from those concepts as describe in detail above. Further conclusions for supply chain design and management can be drawn from nature.

Three categories of critical success factors have been differentiated: technology, cooperation (communication, trust, ...) and complexity.

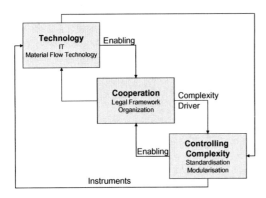

Fig. 2. CSF categories.

Technology related CSF are enabling cooperation. Both – technology and cooperation characteristics – are influencing complexity related issues. The better complexity can be controlled, using appropriate instruments, the less critical the other categories.

8 REFERENCES

[1] Barzel, Y. (1997), *Economic Analysis of Property Rights*, Cambridge University Press, Cambridge.

[2] Bliss, C. (2000), *Management von Komplexität. Ein integrierter systemtheoretischer Ansatz zur Komplexitätsreduktion*, Gabler, Wiesbaden.

[3] Christopher, M. and Peck, H. (2004), *Building the Resilient Supply Chain*, Cranfield School of Management, Cranfield.

[4] Coase, R.H. (1937), *The Nature of the Firm*, Economica, No. 4, pp. 386-405 [http://staff.bath.ac.uk/mnsbr/mang0094/Coase(1937).pdf, 13/07/2006].

[5] Cooter, R. and Ulen, Th. (1988), *Law and Economics*, Harper Collins Publishers, NY.

[6] Dieckmann, U., Doebeliand, M. and Metz, J.A. (2004), *Adaptive Speciation. Cambridge Studies in Adaptive Dynamics*, Cambridge University Press, Cambridge.

[7] Dove, R. (1995), *Best Agile Practice Reference Base - 1994: Challenge Models and Benchmarks*, Proceedings: 4th Annual Agility Conference, Agility Forum, Bethlehem, PA, [http://www.parshift.com, 15/07/2006].

[8] Eccles, R.G. (1981), *The quasi-firm in the construction industry*, Journal of Economic Behavior and Organization, Vol. 2, pp. 335-357.

[9] Engelhardt, C. (2001), *Informationspathologie: Störungen und Ineffizienzen im Informationsfluß einer Organisation*, unpublished doctoral dissertation, University of Leoben, Institut Wirtschafts- und Betriebswissenschaften, Leoben.

[10] Engelhardt-Nowitzki, C. and Zsifkovits, H.E. (2006a), *Open Variant Process Models in Supply Chains*, to be published in: Blecker, Th. and G. Friedrich (2006), *Mass Customization Information Systems in Business*, Idea Group Publishing, Hershey, London (forthcoming).

[11] Engelhardt-Nowitzki, C. and Zsifkovits, H.E. (2006b), *Creating Adaptive Supply Chain Networks*, Proceedings of ICAM 2006 (International Conference on Agile Manufacturing), Norfolk, Virginia, July 2006.

[12] Fama, E.F. and Jensen, M.C. (1983), *Separation of Ownership and Control*, Journal of Law and Economics, Vol. 24, Issue 2, S. 301-326.

[13] Ferger, Ch. (1997), *Die Optimierung von Fließsystemen. Formulierung und Systematisierung von Gestaltungsideen für die Logistik in disziplinärer und interdisziplinärer Umschau*, unpublished doctoral dissertation, University Erlangen-Nürnberg.

[14] Furubotn, E.G. and Richter, R. (1997), *Institutions and Economic Theory: The Contribution of the New Institutional Economics*, University of Michigan Press, Ann Arbor.

[15] Haberfellner, R. and de Weck, O. (2005), *Neuere Entwicklungen im Systems Engineering – Agile SYSTEMS ENGINEERING vs. AGILE SYSTEMS Engineering*, in: Engelhardt-Nowitzki, C. and Wolfbauer, J. (2005), *Gelebtes Netzwerkmanagement, Festschrift für Albert F. Oberhofer zum 80. Geburtstag*, Göttingen, pp. 129-149.

[16] Horvath, L. (2001), *Collaboration: key to value creation in supply chain management*, Supply Chain Management: An International Journal, Vol. 6 No. 5, pp. 205-207.

[17] Hoogervorst, J.A.P., Koopman, P.L. and van der Flier, H. (2005), *Total quality management The need for an employee-centred, coherent approach*, in: The TQM Magazine Vol. 17 No. 1, 2005, pp. 92-106.

[18] Huotari, M.-L. and Wilson, T.D. (2006), *Determining organizational information needs: the Critical Success Factors approach*, Information

[19] Jarillo, J.C. (1988), *On strategic networks*, Strategic Management Journal, Vol. 9, 1/1988, pp. 31-41.

[20] Jarillo, J.C. (1993), *Strategic networks: Creating the borderless organization*, Butterworth-Heinemann, Oxford.

[21] Jensen, M.C. and Meckling, W.H. (1976), *Theory of the Firm: Managerial Behavior, Agency Costs and Ownership Structure*, Journal of Financial Economics, Vol. III, pp. 305-360.

[22] Kaluza, B., Dullnig, H. and Malle, F. (2003), *Principal Agent-Probleme in der Supply Chain – Problemanalyse und Diskussion von Lösungsvorschlägen*, Diskussionsbeiträge des Instituts für Wirtschaftswissenschaften der Universität Klagenfurt, Nr. 2003/03, Universität Klagenfurt.

[23] Klaus, P. (1994), *Jenseits der Funktionenlogistik: der Prozeßansatz*, in: Isermann, H. (1994), *Beschaffung - Produktion – Distribution*, Landsberg / Lech, pp. 331-348.

[24] Libecap, G.D. (1999), *Contracting for Property Rights*, University of Arizona, Tucson and National Bureau of Economic Research, Cambridge.

[25] Mason-Jones, R. and Towill, D. (1998), *Shrinking the Supply Chain Uncertainty Cycle*, Control, September 1998, pp 17-22.

[26] Mentzer, J.T., Foggin J.H. and Golicic S.L. (2000), *Collaboration: The Enablers, Impediments, and Benefits*, Supply Chain Management Review, Vol. 4 No. 5, pp. 52-61.

[27] Milgrom, P. and Roberts, J. (1992), *Economics, organization and management*, Prentice-Hall, Englewood Cliffs, N.J.

[28] Mumenthaler, M. and Mattle, H. (2006), *Fundamentals of Neurology: An Illustrated Guide*, Thieme, Medical Pub, Stuttgart.

[29] Nicita, A., Rizzolli, M. and Rossi M.A. (2005), *Towards a Theory of Incomplete Property Rights*, ISNIE Conference, Barcelona, 22-25 Sept. 2005 [http://www.jus.unitn.it/cardozo/Review/2005/Property.pdf, 13/07/2005].

[30] Norman, A.W. and Litwack, G. (1997), *Hormones*, Academic Press, San Diego, CA.

[31] Ouchi, W.G. (1980), *Markets, bureaucracies and clans*, Administrative, Science Quarterly, Vol 25, pp. 129-141.

[32] Picot, A. and Böhme M. (1999), *Controlling in dezentralen Unternehmensstrukturen*, Vahlen, München.

[33] Pfohl, H.-Ch. (2003), *Logistiksysteme. Betriebswirtschaftliche Grundlagen*, 7. Auflage, Springer, Berlin et al.

[34] Poirier, Ch.C. and Quinn, F.J. (2003), *A Survey of Supply Chain Progress*, Supply Chain Management Review, Vol. 7, No. 5, pp. 40-47.

[35] Raduescu, C. and Heales, J. (2004), *The Role of Management Incentives in Successful Information Systems Development and Implementatio,*. Working Paper, University of Queensland [http://www.business.uq.edu.au/events/speakers/heales-raduescu-paper.pdf, 15/07/2006].

[36] Rockart, J.F. (1979), *Chief executives define their own data needs*, Harvard Business Review, 57(2), pp. 81-93.

[37] Rosenau, J.N., Smith, St., Biersteker, Th., and Brown, Ch. (1997), *Along the Domestic-Foreign Frontier: Exploring Governance in a Turbulent World*, Cambridge University Press, Cambridge.

[38] Russell, S. and Norvig, P. (1995), *Artificial Intelligence: A Modern Approach*, Prentice-Hall, London, pp. 31.

[39] Sappington, D.E.M. (1991), *Incentives in Principal-Agent Relationships*, Journal of economic perspective, Vol. 5, Issue 2, S. 45-66.

[40] Sauvee, L. (2002), *Efficiency, Effectiveness and the Design of Network Governance*, 5th International Conference on Chain Management in Agribusiness and the Food Industry, Noordwijk an Zee, The Netherlands, June 7-8, 2002 [http://www.crepa.dauphine.fr/ArticleCahierRecherche/Articles/LoicSauvee/ICCM%20Noordwijk-2002.pdf, 08/04/2006].

[41] Simatupang, T.M. and Sridharan, R. (2002), *The collaborative supply chain*, International Journal of Logistics Management, Vol. 13 No.1, pp. 15-30.

[42] Spath, D., Wagner, K., Aslanidis, S., Banner, M., Rogowski, T. Paukert, M., and Ardilio A. (2006), *Die Innovationsfähigkeit des Unternehmens gezielt steigern*, in: Fokus Innovation, editor: Bullinger, H.-J., Hanser, München.

[43] Watts, R. and Zimmerman, J. (1986), *Positive Accounting Theory*, Prentice-Hall, Englewood Cliffs, N.J.

[44] Weiss, Th.G. (2000), *Editor of Global Governance: A Review of Multilateralism and International Organizations*, No. 5/2000.
[45] Williamson, O.E. (1996a), *The economic institutions of capitalism: Firms, market relational contracting,* New York.
[46] Williamson, O.E. (1996b), *The Mechanisms of Governance,* Oxford University Press, Oxford.

The Steyr Network Model as a New Management Approach for Modern Logistics

Evelyn Rohrhofer[a], Karin Spaeth[a]

[a]Upper Austria University of Applied Sciences, Campus Steyr,
Wehrgrabengasse 1-3, A-4400 Steyr, AUSTRIA

ABSTRACT

Motivation: The definition of logistics changed considerably over the last years. Thus, today's companies have to holistically manage their multifaceted environment in the sense of supply chain management (SCM) and value network management (VNM). In addition, the incorporated logistics requirements will even expand and enhance in the future. In order to deal with the growing level of intricacy an integrated management model is essential.

Results: As investigations show none of the existing management approaches is applicable in the sense that it is able to entirely fulfill the requirements of SCM and VNM. To eliminate present deficiencies the *Logistikum - Competence Center Logistics and Enterprise Networks* established the Steyr Network Model (SNM). The SNM is distinguished as a holistic, integral management model which supports companies, helping them to define, develop, realize, practice and evaluate successful partnerships in supply chains and value networks.

Availability: The SNM assists companies to manage supply chains and value networks holistically and successfully. By the integrated consideration of the essential management functions companies can effectively improve their operating result.

Contact: evelyn.rohrhofer@logistikum.at
karin.spaeth@logistikum.at

1 INTRODUCTION

Today's market challenges force companies to constantly develop new ways of creating and delivering customer value. As a result of the effects of global competition, the growing power of the buyer as well as the constantly increasing levels of process dynamics and complexity, logistics plays a decisive role in order to reliably fulfill consumer needs. Thus, in the same way as we have seen enormous changes in market conditions in the past decades the focus of activities in logistics has also expanded considerably. Today logistics is an integral economic factor of success and center of interest in management practice [see e.g. Corsten and Gabriel, 2004; Klaus, 2002].

Despite this growing proliferation and relevance of logistics both in theory and practice there is still no uniform answer to the question: "What is logistics?" [Pfohl, 2000]. As a result of the scarcity in literature and the above-named developments in business the *Competence Center Logistics and Enterprise Networks* at the Upper Austrian University of Applied Sciences, Campus Steyr, was introduced to establish the following appropriate understanding of logistics as a basis for all scientific and entrepreneurial activities: Logistics is the end-to-end planning, designing, steering and optimizing of internal and external flows of goods including the corresponding flows of information.

With the increasing levels of complexity and integration four different developments of logistics can be distinguished:

(1) logistics as transportation, handling and storage (basic physical activities of logistics)
(2) logistics as internal coordination (integral logistic coordination and optimization of the internal functional divisions planning, sourcing, production and distribution)
(3) supply chain management (planning, steering and optimizing of supply chains, as defined intra- and cross-company material and corresponding information flows are bundled to conjointly serve uniform customer and process requirements)
(4) value network management (integrated categorizing, planning, steering and optimizing of all defined (supply chains) and undefined (common business connections) material and corresponding information flows which create customer value)

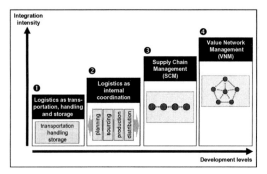

Fig. 1. The four development and integration levels of logistics.

Subject to the displayed economic challenges of today efficient cross-company cooperation has become the key element of competitiveness [Corsten and Gabriel, 2004]. Accordingly, more and more managers have to cope with complex and dynamic supply chains and value networks and find new ways to optimize business processes with partners up- and downstream. In this context logistics in the modern meaning of supply chain management (SCM) and value network management (VNM) has evolved to a new management approach.

Potentials related to SCM are analyzed and verified in several studies and research works which impressively demonstrate success stories of excellent supply chain management. Thus, D'Avanzo et al. highlight the direct link between successful SCM and financial performance [D'Avanzo et al., 2003]. A study of AMR Research identified the top 25 in SCM and declared Dell, Nokia and Procter & Gamble as the leading SCM-performers – the economic success of these companies is indisputable [Friscia et al., 2004].

Despite these resounding potentials a study at the *Competence Center Logistics and Enterprise Networks* shows some unequal results [Losbichler et al., 2006]. The survey which analyzed the European cash-to-cash (C2C) cycle performance as one of the key value drivers for SCM showed two major outcomes:

- The analysis does not confirm the trend of companies like Nokia which were able to benefit from SCM and reduced their C2C cycle extensively. In contrast, the study on European level identified only insignificant enhancements concerning the C2C cycle (from 56 days in 1995 to 54 days in 2004 on average).

- Owing to the considerable success of particular companies and industry sectors on the one hand (e.g. OEM) and the remarkable negative trends in certain branches and companies on the other hand (e.g. subcontractors) the study raises the question whether there are companies or industries which reduce their C2C cycle at the expense of others (e.g. automotive industry, food trade, furniture industry).

To summarize the statements of the paragraphs above: Despite repeatedly quoted success stories the theoretical and practical investigations of cross-company management in the sense of SCM and VNM has shown that there is still a lot of potential in managing and optimizing material flows to the advantage of all supply chain partners. Thus, the question is why some companies are more successful in SCM than others. From the authors' point of view many companies fail because they do not know-how to deal with the complexity of company networks. They voice the need for more and better information about the goals, procedures and methods of SCM [D'Avanzo et al., 2003]. In order to overcome this lack of know-how a holistic management model for supply chains and value networks is essential. This model needs to be an integral approach which offers every company the appropriate framework for successful implementation and long-term logistics performance independent of industrial affiliation, company size and the stage in the process in value creation. The evaluation of established management models and the development of a new approach which is especially focused on the requirements of SCM and VNM are described below.

2 VERIFICATION OF EXISTING MANAGEMENT APPROACHES ON MODERN LOGISTICS REQUIREMENTS

To develop a management model with special regard to the needs of modern logistics, initially the requirements for SCM and VNM have to be identified. Based on the detected needs existing approaches can be verified. The conclusions then are the basis to develop a qualified and suitable management model.

2.1 Requirements for a management model regarding SCM and VNM

A management model designed especially to support the steering and optimizing of supply chains and value networks has to meet the following special requirements arising from the consideration of modern logistics [Staberhofer and Wallner, 2006]:

(1) **Customer orientation:** SCM focuses on the fulfillment of customer needs [Pfohl, 2000]. Therefore all activities within the supply chain aim at granting customer orientation by meeting product, logistics and service criteria and simultaneously assuring ideal efficiency [Christopher, 1998].

(2) **Process perspective:** Product development, order processing, production planning, purchasing, production, distribution and disposal are the necessary processes within the value networks to achieve satisfied customers [Thaler, 2001]. Following the idea of SCM processes have to be considered in a comprehensive and integrated way.

(3) **Mutability:** The permanently growing environmental challenges force companies to create a unique output in the shortest time possible. In this context mutability evolves to a crucial factor to successfully differentiate a company from competitors [Drucker, 1998].

(4) **Cross-company networking:** Supply chain processes do not stop at a company's border but include all supply chain partners. Therefore the consideration of the company's environment and network relationships is essential.

In addition to the mentioned requirements concerning SCM and VNM a management model needs to fulfill some general criteria to enable a successful application or implementation on the part of the user:

- Pragmatism: A management model has to reduce complexity without falsifying. Coherences have to be shown clearly. Issues have to be made visible to the user.
- Feasibility: The usage of management models enables the deduction of actions and improvements.
- Flexibility: Management models are characterized by universality and adaptability to different environments.

- Holism: The model grants interdisciplinary and integrated solutions both for a company's internal and external vicinity [Eschenbach and Kunesch, 1996].

2.2 Analysis of existing management concepts and their suitability for SCM and VNM

In order to deal with the increasing challenges in business different management approaches and concepts have emerged in the last decades, whereby none of these approaches have been able to achieve clear supremacy [Eschenbach and Kunesch, 1996]. Due to the different objectives and directions the evaluation of offered management models – as far as it is possible using public sources – is a complex and almost unfeasible task. Nevertheless the primary comparison of existing approaches with the requirements of SCM and VNM shows the following remarkable deficiencies:

- So far management concepts concentrate on the introversive perspective [Funck, 2001].
- Management concepts do not sufficiently consider the growing significance of cross-company cooperation and neglect the supply chain and value network approach [Funck, 2001].
- The majority of existing management concepts focus on specific issues, e.g. quality management and environment management.
- Many management approaches do not consider all economic aspects and interdependencies of a company and therefore do not fulfill the requirements of an integral and holistic management model. They focus on special issues or divisions and thus neglect others. For instance, the management approach of Drucker has its strengths in offering a framework for strategic planning and development but does not give detailed support for cultural, process-, controlling- or structure-related progress [Eschenbach and Kunesch, 1996]. In contrast, the Balanced Scorecard by Cooper and Kaplan focuses on the strategic aspects of controlling and does not pay enough regard to the optimization of structures and processes.
- Some management approaches have a high level of abstraction and a lack of tools, instruments and methods which hampers concrete realization. In this connection e.g. the approaches of Drucker, Porter and Malik are mentioned in literature [Eschenbach and Kunesch, 1996].
- Another weakness is the bad practicability and adaptability of some approaches. Thus many management concepts are only suitable for companies of a certain branch or a certain size, e.g. Turnheim or Pümpin [Eschenbach and Kunesch, 1996].

Compared to the above mentioned concepts and their inapplicability for SCM and VNM the holistic management approach from St. Gallen appears to be a suitable starting point for further examination. The model by St. Gallen which was developed by Ulrich in the 1970ies can be designated as the basic concept of economics and is – at least in the German-speaking region – a popular and wide-spread approach [Eschenbach and Kunesch, 1996]. Therefore the authors now concentrate on a closer look at the management model by St. Gallen as far as it is described and understood in relevant literature.

The management concept by St. Gallen converts the system-oriented business economics in a model designed for executives and business management [Ulrich, 2001]. By checking this management approach against the supply chain requirements mentioned above some strength and weaknesses can be identified.

- *Customer Orientation*: The model is designed to support executives' decisions when setting up their market service on the basis of customer segments and needs. However, a clear focus of the company's value-creating activities on the customer is not apparent.
- *Process Perspective*: Processes are considered but mainly in an intra-corporate context. Within SCM the comprehensive cross-

company processes possess at least the same significance.

- *Mutability*: The model includes mutability by distinguishing between repetitive and innovative tasks. Basically it is necessary to decide between maintaining an establishment (repetitive task) or substituting an establishment with something new due to environmental changes (innovative task). Moreover, mutability is granted by the organizational level of the model.

- *Cross-company networking*: Environmental conditions are the basis of the company's behavior and form a company's network. Examining common literature the reader gets the impression that the model by St. Gallen solely includes business connections in straight direction to the purchasing and selling market. Following the idea of SCM also e.g. tier 2 and tier 3 are part of the consideration.

- *Other distinctive features*:

Feasibility: The model by St. Gallen is described primarily as a framework for executives of all management levels to basically understand a company's interdependencies regardless of the company's size or industrial sector [Eschenbach and Kunesch, 1996]. For SCM it is furthermore necessary to get recommendations for the execution of adequate tools and methods regardless the user's status.

Holism: The criteria of holism is basically fulfilled as the approach by St. Gallen maps a company as the interaction of the management disciplines strategy, structure, processes and culture. However, according to the saying "if you can't measure it you can't manage it" controlling aspects are more or less disregarded and outside the focus.

Weaknesses of the management model by St. Gallen related to SCM and VNM	Conclusion for SNM
Theoretical focus	Combination of theory and operational practicability
Lack of explicit concentration on logistics to entirely fulfill customer requirements	Focus on all processes which deliver customer value
Company network is limited to immediate environment	Considering network relations regarding the level in the SC
Concentration on internal processes	Extension to comprehensive and cross-company processes
Focus on decision support for executives	applicable to all employees
Secondary and not explicit role of controlling	Controlling has the same importance as the other management disciplines
No concrete activity recommendation	Inclusion of measures and tools for transfer into operational business

Table 1. Comparison of the management model by St. Gallen with the demands for the SNM.

3 A NEW MANAGEMENT APPROACH FOR SCM AND VNM: THE STEYR NETWORK MODEL

Based on the identified requirements of modern logistics and the detected deficits of the approaches mentioned above the *Competence Center Logistics and Enterprise Networks* developed a new management model. The Steyr Network Model (SNM) is distinguished as a holistic, integral management approach which supports companies to define, develop, realize and evaluate successful partnerships in supply chains and value networks. The SNM aims at achieving an optimum in outcome through proactive and reactive customer orientation in the whole network. Furthermore the SNM can be characterized as follows:

- Integration of all significant management functions to enable the development and management of concrete, holistic network solutions

- Contribution of tangible recommendations based on a systematized procedure which support and guide companies by establishing supply chain partnerships and grant re-

sults with sustainable relevance for operational business
- Due to the design on low and high complexity level customized solutions are achievable
- Focus on practicability by development of procedures and tools for the purposeful establishment of relationships, always considering the individuality of the company when acting in cross-company networks
- Supply chain agreements to define the goals, duties and responsibilities between the partners as indicators for the quality of the network collaboration
- Broad field of practical application due to comprehensibility, adaptability and independency from special industries

As stated above, a major characteristic of the SNM is the focus on two different complexity levels in order to achieve the goal – integration of companies to grant results with relevance for operational business – in a smooth way:

- *Low complexity level*: On a low complexity level the foundation for a common supply chain understanding is created. As optimization activities are limited to a product or department view concrete results can be achieved on a short-time basis. Thus, this level especially enables SME the efficient development of network relationships.
- *High complexity level*: A long-term focus on SCM and VNM goes beyond a partial view and affords the interaction with high complexity of cross-company businesses. On this level the central point is the holistic creation and management of value networks including all its participants.

As previously mentioned the SNM is characterized by the comprehensive integration of all fundamental management functions. The management functions of the SNM are divided in five elements (strategy, structure, processes, culture, controlling) and three supporters (environment & infrastructure, optimization, information & technology).

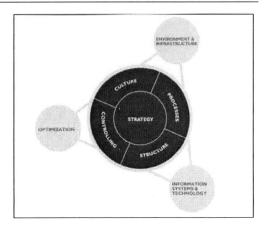

Fig. 2. The design of the Steyr Network Model.

Both elements and supporters accumulate tangible solutions and concepts to ensure the performance of supply chain requirements and the even profit allocation in the network. Thus, each division can be seen as a box of proposals and recommendations which convert into the supply chain agreement.

Subsequently the elements and supporters of the SNM are described in detail.

Element 1: Strategy

The successful strategic positioning in supply chains and value networks requires the development of a cooperative strategy which is clearly aligned to customer needs. In this context the logistics requirements (e.g. delivery times, reactivity, availability or flexibility) are becoming increasingly an essential competitive advantage. Still, in reality supply chain projects are often not adequately aligned with corporate strategy and therefore lag behind the anticipated success. Thus, it is necessary to identify customers' needs, communicate them clearly to the supply chain partners and develop customer-specific supply chain solutions. The comprehensive knowledge of customer requirements then enables the conformable proactive and reactive configuration of the company's processes, structures and culture.

As existing approaches do not sufficiently integrate supply chain management in the strategic planning process the *Competence Center of Logistics and Enterprise Networks* suggests the following procedure in four steps:

- customer analysis (using "requirement radars" and "capability profiles")
- customer segmentation (based on segmentation-technique of Treacy and Wiersema supplemented by logistics requirements [Treacy and Wiersema, 1997])
- strategic positioning (based on criteria of segment attractiveness and "strategic fit")
- supply chain formation (according to customer groups with similar requirements)

Element 2: Structure

Next to corporate strategy the requirements of supply chain management have to be integrated in the organizational structure. For successful supply chains and value networks it is essential to create structures which accurately define and rule the tasks, coordination, communication and responsibilities between all involved partners. Therefore the requirements of customer orientation, process perspective, mutability and cross-company networking have to be achieved both in internal organization and external network structure. Hence the SNM distinguishes between two levels, namely organization and networks.

Organization: The supply chain requirements have to be systematically adapted to build up recommendations for organizational configuration. In this regard the focal points are: segmentation of the organization based on logistic customer requirements, clear process alignment, decentralization of organizational units and development of network capability via modularization [Staberhofer and Wallner, 2006].

Networks: Despite the increasing offers of theoretical and practical insights there is still a great demand in the area concerning which form of network is effective. Thus, a methodology for decision-support to develop successful supply chains and value networks is necessary.

Element 3: Processes

Cross-company process management which is clearly geared to supply chain strategy and customer requirements is an indispensable prerequisite for successful SCM and VNM. However, many companies still show deficiencies in the range of effective configuration, steering and controlling of processes. Correspondingly the process division of the SNM deals with the effective planning, configuration and optimizing of company-internal and cross-company processes according to supply chain requirements. In this regard the lack of uniform standards makes this task nearly impossible. Only if standardized processes are developed and implemented can a supply chain partnership be effective and successful.

In order to make a contribution to the tasks and challenges of process management the *Competence Center Logistics and Enterprise Networks* developed the process reference model LogWIN$_\circledR$-P in cooperation with a large number of companies. The model in four levels accumulates the whole value chain including the processes order management, planning, sourcing, production, distribution and R&D. In this context LogWIN$_\circledR$-P supports companies in the identification, description, optimization and benchmarking of processes by providing a standardized, continuous framework for process management.

Element 4: Culture

Many supply chain projects fail because of deficient acceptance on the part of employees. As supply chains are networks in which various companies work together the diversity of different cultures and values has to be managed adequately. Only if a common supply chain and network culture is developed and anchored in every single company can the potentials of SCM and VNM be fully achieved.

Element 5: Controlling

Due to the rising significance of logistics the measurement of the value contributed to a company's profit gains in importance. The prerequisite therefore is the availability of methods for the planning, steering and controlling of logistics projects. This demands more than efficiency-based performance figures like stocks or cycle times. Since in the modern understanding of logistics the value of a product is created by collaboration within company networks controlling instruments have to be able to quantify and measure the value contribution throughout the whole supply chain. At present no adequate controlling system exists. Furthermore the translation of the value proposition into performance figures is needed. This is the premise for a goal- and value-oriented deployment of logistics activities in the sense of SCM and VNM [Losbichler et al., 2006].

To transact the activities of the five SNM-elements successfully the necessary background in terms of the three supporters described below has to be created.

Supporter 1: Optimization
The multiplicity of logistics tools, methods and concepts has to be adequately converted to the company's demand.

Supporter 2: Environment & Infrastructure
Supply chains and value networks require a perfect infrastructure, common information platforms and the efficient interaction of environmental forces.

Supporter 3: Information Systems & Technology
To create efficient network relationships the accurate adoption of integral concepts in the form of information systems and logistics technologies is essential.

4 CONCLUSION

This paper shows the exigency and significance of a holistic management approach in order to deal with the growing complexity in today's businesses. Since no entirely satisfying solutions could be found in literature a new management model – the Steyr Network Model (SNM) – was established. Recapitulating the SNM can be characterized as an integrated management approach which provides companies regardless of industry and size a holistic framework for creating and optimizing supply chains and value networks. The model which consists of five elements and three supporters is an integral approach which offers a universal structure for practicable, customer-oriented and cross-company management duties in logistics networks.

Despite the established framework and the already developed methods and concepts a lot of research work still has to be done. These research activities especially concern the development of standardized tools to fill and complete the elements and supporters of the model. Furthermore the actual improvements which can be achieved by using the SNM have to be empirically tested.

5 REFERENCES

[1] Christopher, M. (1998) *Logistics and Supply Chain Management*, London.

[2] Corsten, D. and Gabriel C. (2004) *Supply Chain Management erfolgreich umsetzen*, Berlin.

[3] D'Avanzo, R., Von Lewinski, H. and Wassenhove, L. (2003) *The Link between Supply Chain and Financial Performance*, in: Supply Chain Management Review, Vol. 7 No. 11-12.

[4] Drucker, P. (1998) *Die Zukunft bewältigen*, Düsseldorf.

[5] Eschenbach, R. and Kunesch, H. (1996) *Strategische Konzepte, Management-Ansätze von Ansoff bis Ulrich*, Stuttgart.

[6] Friscia, T., O'Marah, K. and Souza, K. (2004) *The AMR Research Supply Chain Top 25 and the New Trillion-Dollar Opportunity*, www.amrresearch.com.

[7] Funck, D. (2001) *Integrierte Managementsysteme*, in: WiSt No 8.

[8] Klaus, P. (2002) *Die dritte Bedeutung der Logistik*, Hamburg.

[9] Losbichler, H., Rothböck, M. and Staberhofer, H. (2006) *Increasing Shareholder Value through Supply Chain Management*, in Pawar, K.S., Yu,

M., Zhao, X., Lalwani, C.S. Competitive Advantage through Global Supply Chains, Proceedings of the 11th International Symposium on Logistics.
[10] Pfohl, H. (2000) *Supply Chain Management: Logistik plus?*, Berlin.
[11] Staberhofer, F. and Wallner, T. (2006) *Supply Chain-orientierte Unternehmensorganisation*, in: Supply Chain Management II/2006.
[12] Thaler, K. (2001) *Supply Chain Management*, Köln.
[13] Treacy, M. and Wiersema, F. (1997) *Markführerschaft: Wege zur Spitze*, Frankfurt.
[14] Ulrich, H. (2001) *Das St. Galler Management-Modell*, in: Gesammelte Schriften Band 2, Bern.

Workshop:

Operations Management

Chairman: Herbert Jodlbauer

Scientific Board:
Manfred Gronalt, Universität für Bodenkultur Wien
Werner Jammernegg, Wirtschaftsuniversität Wien
Herbert Jodlbauer, FH OÖ Campus Steyr

The Concept of "Customer Driven Production Planning" for Evaluation of Job-Shop System Performance

Klaus Altendorfer[a*], Herbert Jodlbauer[b]

[a]Profactor Produktionsforschungs GmbH, Im Stadtgut A2, A-4407 Steyr-Gleink, AUSTRIA
[b]Upper Austria University of Applied Sciences, Wehrgrabengasse 1-3, A-4400 Steyr, AUSTRIA

ABSTRACT

Motivation: The decision to organize production according to "make to stock" (MTS) or "make to order" (MTO) principles is important since it influences finished goods inventory enormously. To evaluate if a production system is able to be run under MTO with zero percentage late jobs the concept presented by Jodlbauer (2006) in "Customer driven production planning" is used. Further some additional information about the performance of a production system which is not able to install a pure MTO system is derived from Jodlbauer`s work. This additional information is the minimum percentage of late jobs which could theoretically be reached by the production system if it is run under MTO, and how much additional capacity would be needed to be able to install an MTO system with zero percentage late jobs.

For a specified job-shop production which is, according to the evaluation based on Jodlbauer, not able to install an MTO system, the minimum percentages of late jobs for different levels of average due dates are calculated corresponding to the methodology presented in this paper. For these levels of average due dates used to calculate the delivery reliability, simulation runs are performed and the simulation results are compared to the results according to the presented methodology. Jodlbauer`s paper focuses on flow-shop productions with low variations on the shop floor and predefined paths for the products through the production. The simulated job-shop production in this investigation is characterized by flexible paths of the products through the shop floor and has high variations in the time for production as well as many disturbances through machine breakdowns. The simulation model of the job-shop production was created as part of the master thesis by one of the authors.

Results: The analytical evaluation, if the specified job-shop production system is able to work under MTO with zero percentage of late jobs, according to Jodlbauer`s methodology is proven to work well, also the evaluated production system is more complex than the production system Jodlbauer based his work on. The additional information about system performance, which is the minimum percentage of late jobs, derived from Jodlbauer`s paper proves to be a good estimator for this number. The evaluated system with the production planning and control (PPC) parameters used for simulation shows a much lower performance.

Contact: klaus.altendorfer@fh-steyr.at

1 INTRODUCTION

The most important target of each company is to generate profit in a sustainable way and this is independent of the industry in which the company is active. For companies which produce goods the ability to generate profit depends very much on the ability to fulfil customer demand within a competitive delivery lead time and without late jobs (Jodlbauer, 2006). To keep the delivery lead time for the customers at an acceptable level an MTS strategy which keeps a defined number of pieces of each product produced on stock can be performed. This creates a very short delivery lead time for the customer, but for the company it presents several disadvantages. First the company has automatically a high level of finished goods inventory (FGI), which costs money. Second if a customer wants his product with a special feature the company is not able to keep the short delivery times, and if some changes in the product have to be implemented, the stock of this special product has to be scrapped. A further disadvantage can be found if the product spectrum of a company is very broad, because then the FGI becomes enormous.

Another way to serve the customer is to perform an MTO strategy and produce only what the customer demands. Then all the problems discussed in the section above disappear. The question for each manufacturing company is, from this point of view, if it is able to implement an MTO system and how to configure the PPC system to implement it.

The following paper presents a solution for the first question and gives some hints how to find a solution for the second one. A methodology is presented which evaluates if a production system is able to utilize an MTO strategy. If it is not able to implement this, the percentage of jobs which will be late if the system is nevertheless run under MTO will be presented. Further, the following paper shows how to determine which capacities need to be increased to enable the system to run under MTO.

2 CONCEPT OF CUSTOMER DRIVEN PRODUCTION PLANNING

Jodlbauer presents in his paper "Customer driven production planning" the following methodology to evaluate if a production system is able to perform an MTO strategy (Jodlbauer, 2006).

It is clear that in a production system which has no or very little FGI, the production lead time has to be shorter than the delivery lead time from the customer. Improvements in factory layout, processes, organisation and the PPC system can lead to a reduction of the work in process (WIP) and further according to Little's Law (Little, 1961) to a reduction of the production lead time.

2.1 The capacity view

The needed capacity of a machine for producing a final product depends on the bill of material, the routing data, the lot sizes and the standard processing times. The average capacity needed by one machine to produce all the products for one time period can be calculated according to the average product mix. The demand of the market fluctuates which means that in some periods more capacity on a machine would be needed, in others there is some free capacity. To balance these fluctuations the machine can produce some products in time periods with free capacity which will be needed in later time periods. How far into the future the demand should be produced on one machine can be calculated based on the level of fluctuations the demand has. This time period is called capacity work-ahead-window (WAW) h_k for machine k. The capacity WAW for a machine is, according to Jodlbauer (2006), the number of time periods within which the average capacity needed is 99% of the time lower or equal the maximum capacity of the machine. The following equation calculates the capacity WAW h_k for machine k:

$$h_k = \left(\frac{2.33 * \sigma_k(1)}{\overline{c_k} - \mu_k(1)} \right)^2 \quad (1)$$

Whereby:
- $\sigma_k(1)$ ... standard deviation of capacity needed at machine k related to single time periods
- $\mu_k(1)$ ... average capacity needed at machine k for one time period
- $\overline{c_k}$ ... maximum available capacity of machine k

To create a better understanding of the whole concept an example will be used to show how all the steps described work together. This example consists of a 9 machine job-shop production where 3 machine-groups can be defined and 5 products are produced. Each product has to be processed on each machine-group. All five products have the same routing through the shop, which is machine-group A then B and C at the end. Each machine has a capacity of 16 hours per day, which corresponds to 2 shifts of 8 hours each. In the following Fig. 1 the processing time in minutes is indicated under the product types for machine-group A – B – C.

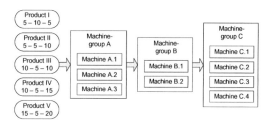

Fig. 1. Example job-shop production.

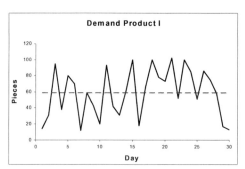

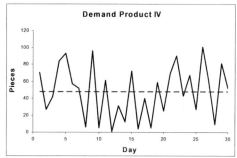

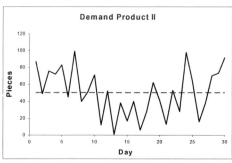

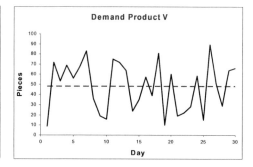

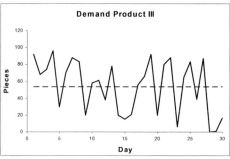

Fig. 2. Demand fluctuations of the 5 products within the last month.

The demand fluctuations of the five products over the last month are shown in Fig. 2. According to Jodlbauer (2006) a good procedure is to take the fluctuations of the last year into account, and use the forecasted demand from the next month as average for the demand fluctuation. In the presented example just the last month has been used, and the demand is approximated to be stable. Fig. 3 shows how the fluctuation of the demand and the average capacity needed performs for the three machine-groups of the example with the capacity WAW calculated according to equation 1. As pictured there the longest capacity WAW h_k emerges with machine-group B.

2.2 The customer view

The statistical distribution of the delivery lead time of a product describes the statistical behaviour of the time which is available for the production system to fulfil the customer orders for this product and is called order characteristic $OC_i(\Delta)$ of product i (Δ denotes the difference between planned delivery date and order date). Each product has a certain minimum necessary production lead time which it needs for technical reasons. This is the time for processing and transportation, but not the waiting time in front of machines and is called $l_{min}(i)$ for product i. This $l_{min}(i)$ depends further on the size of the production lots, since big lot sizes lead to a greater value than small ones since every produced part has to wait for the whole lot to be processed on a machine. All the demands for product i with a delivery lead time between 0 and $l_{min}(i)$ have to be held on stock in the FGI or will not be delivered on time. The parameter $l_{min}(i)$ is in the presented example 1 day for all products, which means it takes at least 1 day from order entry to delivery of a product. Fig. 4 shows how this order characteristic $OC_i(\Delta)$ looks like for the five products of the presented example.

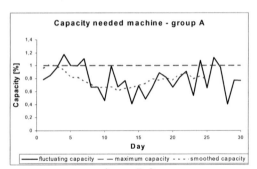

$h_A = 5,8$

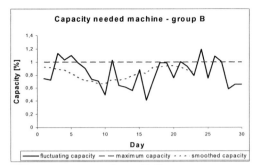

$h_B = 7,7$

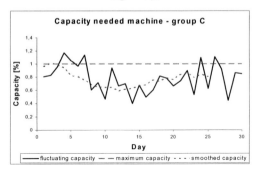

$h_C = 5,4$

Fig. 3. Capacity fluctuation for the three machines.

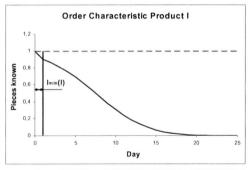
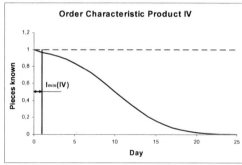
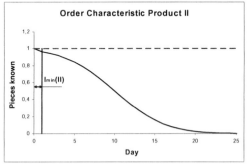
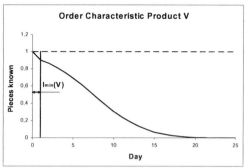
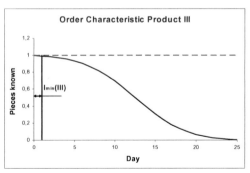

Fig. 4. Order characteristics of the five products.

In this example for product I 80% of all the pieces ordered had a delivery lead time of 3 days or more. This means more than 80% of the demand of this product is known within a time window of 3 days into the future.

Based on all products a machine can produce and on the bills of material, the standard processing times including setups as well as the routing data of all the products an operations characteristic $OCM_k(\Delta)$ can be calculated for each machine or machine-group k. According to the products produced the machines also have a minimum possible lead time l_{min_k}. Fig. 5 shows the operations characteristics for all the machine-groups of the integrated example, whereby all values for l_{min_k} are one day.

For each machine or machine-group the capacity WAW h_k and the operations characteristic $OCM_k(\Delta)$ is now known. To use the capacity WAW of a machine all the demand within l_{min_k} and h_k would have to be known.

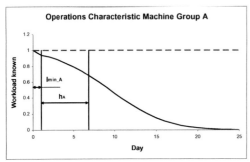

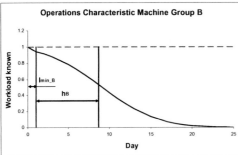

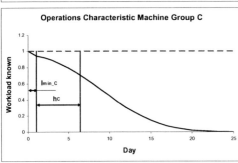

Fig. 5. Operations characteristics of the machines.

According to the operations characteristic it is obvious that not all this demand is known. For this reason the WAW for planning of a machine h_{p_k} has to be enlarged in a way that the unknown demand between l_{min_k} and h_k is covered by orders which have their delivery date further in the future than h_k. The following equation shows how to calculate h_{p_k}:

$$h_{p\_k} - h_k = \int_{l_{min\_k}}^{l_{min\_k}+h_{p\_k}} 1 - OCM_k(\Delta)d\Delta \qquad (2)$$

If it is not possible to calculate h_{p_k} for all machines or machine-groups, then the production system is not able to perform an MTO strategy. So the calculation of h_{p_k} is the check for the production system for its ability to run under MTO.

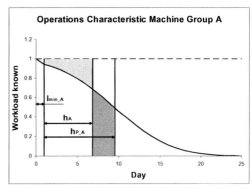

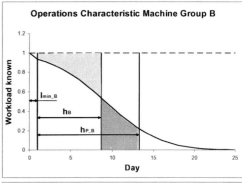

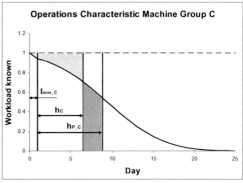

Fig. 6. Operations characteristics with h_{p_k} calculated.

Fig. 6 shows the operations characteristics of the three machine-groups within the presented example. Further the three parameters l_{min_k}, h_k and h_{p_k} for the three machine-groups are given and for each machine the two areas which have to be equal are marked grey. For the presented example machine-group B is the bottleneck and should be able to start jobs whose due date are up to 13,2 days in the future.

2.3 PPC parameters for CONWIP

For each single machine or machine-group it is also possible to calculate the remaining production lead time L_k if h_{p_k} is implemented according to the following equation:

$$L_k = \frac{\int_0^{l_{min}+h_{P\_k}} f(\Delta) * \max(\Delta, l_{min\_k}) d\Delta}{\int_0^{l_{min}+h_{P\_k}} f(\Delta) d\Delta} \quad (3)$$

Whereby:

$$f(\Delta) = \frac{\partial}{\partial \Delta} F_{\mu_{\Delta Mk}, \sigma^2_{\Delta Mk}}(\Delta) \quad (4)$$

And:

$$F_{\mu_{\Delta Mk}, \sigma^2_{\Delta Mk}}(\Delta) = 1 - OCM_k(\Delta) \quad (5)$$

For a production system with a CONWIP PPC-system Jodlbauer (2006) describes how to calculate the parameters WAW and WIP-Cap (WC). Further information about CONWIP can be found in Spearman et al. (1990) and Hopp and Spearman (1990, p. 423 pp). The WAW can be implemented as the maximum of all l_{min_k} plus the maximum of all h_{p_k}. For the presented example this would be $h_{p\_B} + l_{min\_B} = 12{,}2 + 1$ which are 13,2 days. The WC should, according to Jodlbauer (2006), be the maximum of all remaining production times multiplied by the average capacity of this machine or machine-group. In the presented example this would be $L_B * \overline{c_B}$ = 7.6 * (16 hours working time per day * 2 machines) = 243.2 [hours of work for machine-group B].

3 ADDITIONAL SYSTEM PERFORMANCE INFORMATION

In addition to the information as to whether a production system is able to perform an MTO strategy and how the parameters have to be set in this system, further information can be derived from Jodlbauer's work.

3.1 Greatest possible service level

If the analyzed production system is not able to implement a pure MTO strategy, according to the evaluation described in the upper section, but works under MTO, the theoretical maximum percentage of jobs on time, defined as service level for the remainder of the paper, can be calculated.

First the maximum possible value for h_k (called h_{k_max}), depending on the operations characteristic of a machine k, has to be calculated for all machines or machine-groups according to the following equation:

$$\int_0^{h_{k\_max}} 1 - OCM_k(\Delta) d\Delta = \int_{h_{k\_max}}^{\infty} OCM_k(\Delta) d\Delta \quad (6)$$

Second the probability of required capacity being smaller than the maximum available capacity is calculated using h_{k_max}. This probability can be assumed to be equal to the maximum percentage of jobs on time for machine k:

$$T_k = F_{0,1}\left(\frac{\sqrt{h_{k\_max}} * (\overline{c_k} - \mu_k)}{\sigma_k}\right) \quad (7)$$

$$T_{max} = \min(T_k) \quad \text{For k=1 ... n} \quad (8)$$

Whereby:
- T_k ... theoretical maximum percentage of jobs on time for machine or machine-group k
- T_{max} ... theoretical maximum for the service level
- n ... number of machines or machine-groups in the production system
- $F_{0,1}$ ... distribution of the capacity disturbances, normal distribution is assumed for the remainder

The service level which is calculated with equation (8) is not exactly the one which is measured by a company since it is not based on the number of jobs which will be on time. The calculated service level states which amount of work cannot be delivered on time even if all known demand is taken into account. For a usual production system this difference should have just marginal effects. Further it is important that the result of equation (8) is not the expected service level of a production system, but the greatest possible for this system if everything is optimally tuned.

3.2 Changes needed to enable compatibility with MTO

Further information which is of interest for the production manager can also be derived from the work of Jodlbauer (2006).

If the production manager knows which target WAW h_t he wants to implement in his production system, this should be a feasible value, then he has two opportunities to achieve this target. On the one hand it is possible to reduce the variance of the statistical distribution of the customer demand by influencing the sales department. This could result in a minimum delivery lead time of a few days which each customer has to accept. If this is possible and does not impact the sales volumes in a negative way, it could be a good way to improve the overall service level. How much the variance has to be changed can be evaluated by changing it so far, that for all machines or machine-groups max(l_{min_k}) + h_{p_k} becomes smaller or equal to h_t. Since this solution is not very often a feasible way because the customer demand is often unchangeable for the production manager the second way to react is to increase the production capacity. The production capacity of all machines or machine-groups which have a max(l_{min_k}) + h_{p_k} greater than h_t has to be increased until max(l_{min_k}) + h_{p_k} is smaller than or equal to h_t. Fig. 7 shows the two possibilities to make the production system able to run under MTO.

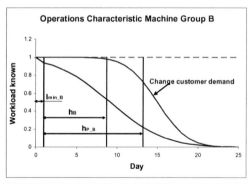

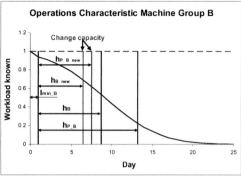

Fig. 7. Possibilities to become able for MTO.

The value of needed capacity increase gives the production manager a good estimator of how far the system is away from being able to perform an MTO strategy.

4 SIMULATION MODEL FOR VERIFICATION

To confirm the finding that it is possible to evaluate the maximum service level for a production system which performs an MTO strategy but is not able to reach a service level of 100% according to equation (2) the following simulation model is used.

The simulated production is the last part of an electronic parts factory where the parts are tested for whether they are valid or invalid. The simulation model was created for the master thesis of Altendorfer (2006, p. 58 pp) and is in this study used to confirm or reject the above findings.

The production is a job-shop with a sequential bill of material. It consists of 33 machines and has about 500 different products which are produced 7 days a week 24 hours a day. Each single product has between 1 and 9 processing steps which can be performed on between 2 and 25 machines. This means flexible routing is possible. The machine for the first process step can also be the one for the last process step. It is not possible to group the machines since every product has for every process step a special set of machines which can perform this process step. The shop load is 85% which includes errors and maintenance. The due dates are calculated according to the following equation similar to Rohleder and Scudder (1993) derived from the total work content method.

$$dd_i = et_i + pt_i * f * (0{,}5 + k * z_{g(0,1)}) \quad (9)$$

Whereby:
- dd_i ... due date of job i
- f ... parameter which defines how long the delivery lead time is on the average
- $z_g(0,1)$ ... random number uniformly distributed between 0 and 1
- k ... constant
- et_i ... entrance time of job i
- pt_i ... sum of all planned processing times of job i

As shown in equation (9) the due date depends on the planned processing time of the job. The production system is operated under 4 levels of average delivery lead time which is implemented by changing the parameter f from 2 to 5. The production system has CONWIP for PPC. For further information about the simulation model see Altendorfer (2006, p. 58 pp).

This evaluated production system is far more complex than the one Jodlbauer (2006) developed his methodology for, but should, for this reason prove, the usability of Jodlbauers work, and the additional information derived in this paper, in the industry.

5 RESULTS

It is not possible to build groups of machines in the simulated production, and it is also not possible to evaluate the operations characteristics for every single machine because each job can, at every process step, be processed on more than one machine. This leads to the situation that the only way to evaluate the production system performance is to build one machine-group including all machines. This machine-group has one operations characteristic which includes the whole job-shop production. For all the four levels of average delivery lead time in the first step the maximum possible value for h_k is calculated according to equation (6) for this single machine-group. In the second step the maximum possible service level according to equations (7) and (8) is calculated. Further, at each level of average delivery lead time a simulation run is performed with the following conditions:
- The service level is calculated on the "exact" time base of the simulation, for the calculation the time base for the service level is days
- The average processing time needed is 28% higher than the average processing time planned, which leads to high fluctuations in processing times
- An average of 14% of the available time is used for disturbances, this value is included in the calculation of h_{k_max}, but leads to a weak system performance

Fig. 8 shows the calculated maximum service level versus the reached service level in the production system.

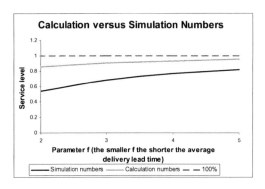

Fig. 8. Calculation versus Simulation numbers.

As shown in Fig. 8 the service level reached within the simulated production system is far below the values calculated for the maximum possible service level. The trend of the production system to yield a lower service level when delivery lead times become shorter is represented very well in the calculation, and also the characteristic of the curve is the same. The big deviation between the two lines can, in our opinion, be explained by the far from optimal conditions within which the simulated job-shop is run.

For a production manager the presented model should give a hint of what is still possible on his shop floor, if the capacities and the demand behaviour stay unchanged. In short, the answer to the question what service level is theoretically possible in the current situation for a production system can be drawn from the presented model.

If a production system is even theoretically not able to implement an MTO system, this means either the capacity has to change, this can be done by an increase of capacity or by some flexible overcapacity, or a certain level of jobs late has to be accepted.

6 CRITICAL ESTIMATIONS WHICH INFLUENCE THE RESULTS

The most important estimation which influences the result is the chosen length of time periods for the evaluation. This influences the fluctuations in demand, which means if the calculation is based on hours the value for h_k will be greater than if it is based on days or shifts. For this reason one has to be aware that the first smoothing of the fluctuations is already made by choosing the time period. Here Jodlbauer (2006) defines as the sensible value the one the service level is based on. This means if the customers measure the service level based on the day a job is finished, the "right" period is a day, if the service level is measured on an hourly basis the "right" value is an hour.

The calculated maximum service level in equation (8) is based on the processing time. This means if the result of equation (8) is 85%, then the maximum reachable service level of the production system is 85% of the work content, these could be 95% of the jobs if there are some very short jobs and others which are very long.

7 CONCLUSION

The concept of Jodlbauer (2006) with the enlargements presented in this paper can be a good tool for evaluating the ability of a production system to work under MTO. It further gives some hints on how far a production system is away from being able to install an MTO strategy and how well it performs in comparison to the theoretical maximum service level. For the production manager a good source of information can be the estimation how much capacity would be needed at which machine-groups to enable MTO with 100% service level, how it is presented in this paper. For the calculation of these values it is important to use the "right" time scale for the time periods within it.

8 REFERENCES

[1] Altendorfer, K.: *Produktionssteuerung mittels Abarbeitungsregeln, Einflüsse von verschiedenen Abarbeitungsregeln innerhalb eines CONWIP Produktionssystems auf den Deckungsbeitrag*, 2006, Steyr.

[2] Hopp, W. J.; Spearman, M. L.: *Factory Physics*, Second Edition, 2000, Boston a. o.

[3] Jodlbauer, H.: *Customer driven production planning*, Working Paper, will be published in, International Journal of Production Research, 2006.

[4] Little, J. D. C.: *A proof for the queuing formula L = λW*, Operations Research, Vol. 9, No. 3, 1961, S 383 – 387.
[5] Rohleder, T. R.; Scudder, G.: *A comparison of order-release and dispatch rules for the dynamic weighted early/tardy problem*, Production and Operations Management, Vol. 2, No. 3, 1993, S 221 – 238.
[6] Spearman, M. L.; Woodruff, D. L.; Hopp, W. J.: CONWIP: *A pull alternative to kanban*, International Journal of Production Research, Vol. 28, No. 5, 1990, S. 879 – 894.

Makespan Minimization for Parallel Machines

Wolfgang Promberger[a], Herbert Jodlbauer[b*]

[a] AT&S China, Jin Du Road 5000, Xinzhuang Industry Park, Minhang District, 201108 Shanghai, P.R.China
[b] Upper Austria University of Applied Sciences Steyr, Wehrgrabengasse 1 – 3, A-4400 Steyr, AUSTRIA

ABSTRACT

Motivation: A model for a production system with multiple parallel machines and product dependent processing times is introduced. The objective is to minimize the maximum capacity needed (= make span) for a product mix structure known in advance.

Results: By applying this model, an optimized loading for each line is calculated, whereby it should be assumed that not every product can be produced on each machine, availability of each line is different, and the product mix is highly variable.

Availability: The introduced model is already applied in an electronics company, and can be applied to any manufacturing company.

Contact: w.promberger@ats.net, herbert.jodlbauer@fh-steyr.at

1 INTRODUCTION

The type of machines to which this model is applied, are so-called horizontal lines: As shown in Fig. 1, all of them consist of three modules, and the products are transported through these modules on a conveyor, whereby the conveyor-speed depends on the product and is different in each module. The products are put on the line automatically with a loading-machine. The time gap between loading two pieces, is different for each product and each line because a minimum gap is required between the products when going through the line.

Moreover, it needs to be considered that:

- not every product can be produced on each machine, and consequently process times are different,
- due to planned downtimes the available capacity of each machine is different, and
- the required product mix is highly variable.

When the required quantity of each product (e.g. demand of one shift) and the availability of each line is known, this model shows the quantity of each product which should be loaded to each line in order to minimize the maximum required capacity of all lines.

Brucker describes basic methods for scheduling of parallel machines. Kang, Mailik and Thomas introduce an interesting method for sequence-dependent setup costs. Moreover, Cambell found a model for short term production planning of parallel machines.

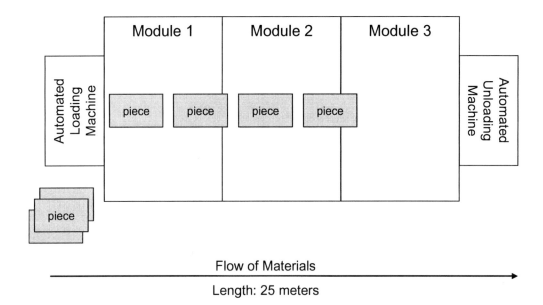

Fig. 1. Layout of one horizontal line.

2 MODEL

A definition of all variables and parameters is as follows:

i ... Index for i-th product
j ... Index for j-th machine
k ... Index for k-th module
Δi ... size of product i in meter

$$a_{i,j} = \begin{cases} 1 & \text{product i can be produced on machine j} \\ 0 & \text{else} \end{cases}$$

c ... required minimum gap between pieces in meter
$l_{j,k}$ ... length of module k of line j in meter
$v_{i,j,k}$ ... conveyor speed of product i throughout module k of machine j in meter/minute
n_i ... lotsize of product i
D_i ... demand of product i in lots
T_j ... maximum available capacity of machine j in minutes
$w_{i,j}$ ... process time for one lot of product i at machine j in minutes
$x_{i,j}$ ... quantity of lots of product i which should be produced on machine j

Due to the minimum required space between two pieces (see Fig. 2), when going through the slowest module of the line, the loading times $t_{i,j}$ can be calculated for each line and product. The formula is based on the Physics Principle that the path covered is the product of time and speed.

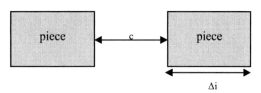

Fig. 2. Gap between two pieces of one lot when going through the slowest module.

$$v_{i,j\min} = \min_k v_{i,j,k}$$

$$t_{i,j} = \frac{c + \Delta i}{v_{i,j,\min}}$$

$v_{i,j\min}$ ... actual conveyor speed for product i at line j in meter/minute

$t_{i,j}$ ... time gap between loading two pieces within one lot of product i at machine j in minutes

In a next step, the process time per lot is calculated for each line: it is the result of the time gap, between loading two pieces on the line, multiplied by lotsize, plus the time required by the first panel to go through the whole line minus its loading time (since there are only n_i-1 gaps between the panels).

$$w_{i,j} = t_{i,j} n_i + \sum_k \frac{l_{j,k}}{v_{i,j,k}} - \frac{c}{v_{i,j\min}}$$

The model consists of 4 constraints:

Demand Fulfillment

$$\sum_j x_{i,j} a_{i,j} \geq D_i$$

Nonnegativity

$$x_{i,j} \geq 0$$

Integer

$$x_{i,j} = \text{integer}$$

Capacity Constraint

$$\sum_i x_{i,j} a_{i,j} w_{i,j} \leq T_j$$

The Demand Fulfillment constraint makes sure that all required products are produced, and the Capacity Restriction that the required capacity of each line is less than or equal to its available capacity.

Target of the model is to minimize the maximum required capacity of all lines.

$$\max_j \sum_i x_{i,j} w_{i,j} \longrightarrow \min$$

The introduced model is an integer linear MIN-MAX optimization problem to search for an optimum $x_{i,j}$. For a real situation with four products, three lines, and three modules in each line, an optimized loading can be calculated with a spreadsheet calculation programme and the tool Solver.

3 RESULT

By applying the introduced model the following goals are achieved:

(1) The maximum required capacity of all machines is minimized.
(2) Utilization of all machines is nearly balanced.
(3) Throughput targets of all products will be met by each department. Therefore, the flow of materials between machines and departments becomes stable and can be planned. Moreover, Work In Process in front of machines can be decreased without any loss in throughput or utilization. Eventually, the lead time of the whole production system will decrease.

4 REFERENCES

[1] Brucker P. (2005) *Scheduling Algorithms*. 2. edition, Springer.
[2] Kang S., Mailik K., Thomas L.J. (1994) *Lotsizing and scheduling on parallel machines with sequence-dependent set up cost*. Working Paper No. 94-96, Cornell University.
[3] Cambell G.M. (1992) *Using short-term dedication for scheduling multiple products on parallel machines*. Production Operations Management 1, 295-307.
[4] Jodlbauer H., Promberger W. (2006) *Minimale Belegungszeit paralleler Anlagen*. PPS Management 1/2006.

New Developments in Scheduling – Improved Model and Solver Approaches

Andreas Weidenhiller [a*], Gabriel Kronberger [b]

[a] Upper Austria University of Applied Sciences, Wehrgrabengasse 1-3, A-4400 Steyr, AUSTRIA

[b] Upper Austria University of Applied Sciences, Hauptstraße 117, A-4232 Hagenberg, AUSTRIA

ABSTRACT

Motivation: This paper presents several conceptual and algorithmic improvements to the multi-item, single-machine dynamic scheduling problem and its solution as recently presented by Jodlbauer.

The dynamic demand which distinguishes the model from the majority of earlier approaches can be modified without changing the position of the optimum, but simplifying the model significantly. Furthermore, one can consistently change over to using cumulated demand in all calculations, thus eliminating several difficulties with discrete demand data. For generalization, sequence-dependent setups can be introduced. By changing from a start algorithm based on the special case of the Economic Production Lot (EPL) to one based on prioritizing product types depending on the value holding cost multiplied by capacity, the quality of the initial solution can be improved considerably.

Results: Comparison with various scheduling problems from the literature yielded a favorable outcome. The improvements in model and solution approach were found to have immediate effects on solution quality.

Contact: andreas.weidenhiller@fh-steyr.at

1 INTRODUCTION

Since Harris' EOQ model (Harris 1913) and the EPL formula developed by Taft (1918), which were some of the first mathematical works on lot-sizing, many kinds of extension and improvement have been suggested, e.g. multi-product and multi-machine environments, capacity restrictions, and various approaches to considering setup activities. For details we recommend the work by Salomon (1991).

Especially the earlier models can be grouped clearly into those using time-discrete approaches on a finite planning horizon and those considering cyclic schedules on an infinite continuous time scale; demand is modeled either as a constant function or as a set of due dates and quantities.

Optimal control theory is also a source of contributions to the field. Holt et al. (1960), Kimemia and Gershwin (1983), and Kogan and Perkins (2003) developed models which all take dynamic demand into account to some degree.

Almost any considered model results at least in a mixed-integer linear program; even moderately complex problems are NP-hard. Consequently, we find solution approaches making use of heuristics such as genetic algorithms, Tabu Search and Simulated Annealing.

Based on ideas of both the discrete models and those from optimal control theory, Jodlbauer developed a new model in (Jodlbauer 2006a), describing a single-machine multi-product capacitated production system for a finite planning horizon. It considers setup times and costs and does not allow backlog. In (Jodlbauer 2006b), Jodlbauer presented a reformulation of his original model along with new concepts concerning the solution of the problem. Jodlbauer and

Weidenhiller (2006) came up with further improvements of the model, especially with a method to abolish the rather complicating notion of Demand Entry Points.

This article first shows some advantages of using cumulative views on demand and production before giving a short introduction to the model conceived by Jodlbauer (2006a). Section 4 summarizes the findings of Jodlbauer and Weidenhiller (2006) as to how the cumulated demand functions can be modified in order to avoid the use of Demand Entry Points while leaving the problem substantially unchanged. A new and effective way of creating a start schedule for our improvement heuristic is presented in Section 5, while Section 6 deals with benchmarks for the solution algorithm. A summary and reflections considering future research conclude the article.

Also at the end, in Table 6, a list of all symbols used can be found for reference.

2 CUMULATIVE VIEWS ON DEMAND AND PRODUCTION

Closer study of the model presented in Jodlbauer (2006a) reveals that cumulated demand and cumulated production output are the relevant entities, not their derivatives, demand rate and production rate.

Aside from this obvious argument for the use of cumulated entities, there are two further advantages: Simpler representation of discrete demand, and more intuitive visualization.

In many practical applications, demand is in fact discrete, that is, a list of due dates and quantities. In the cumulative view, these due dates result in left-sided discontinuities of the function, which can easily be dealt with – as opposed to the introduction of Dirac Delta functions required when working with demand rates.

Looking at a visualization (e.g. Figure 1), due dates can be easily identified because the graph of the cumulated demand "jumps" exactly at these times. The quantity ordered corresponds to the height of the jump.

In graphs where both cumulated production and cumulated demand are presented, backlogs can be identified easily. For a more detailed explanation please refer to the following chapters, an illustration can be found in Figure 3.

Furthermore, we would like to point out that similar cumulative views have been used very successfully in the related field of production monitoring and the throughput diagram (see e.g. Wiendahl and Springer 1988, or Wiendahl 1997).

3 MODEL BACKGROUND

The full mathematical model, which can be found in Jodlbauer (2006a) will not be presented here; instead, we make a quick tour of the key aspects and go into detail only as far as necessary for the requirements of this article.

Figure 1 displays the cumulated demand (D_i, i either A or B) for the two products, A and B, within the time horizon. Clearly, this is a discrete demand structure, where due dates correspond to discontinuities in the cumulated demand.

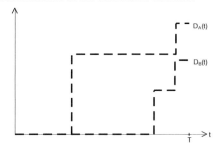

Fig. 1. Discrete demand for two products.

For each product a constant production rate c_i is assumed, and so the cumulated production X_i for each product is piecewise linear, alternating between the rates zero and c_i (see Figure 2). Time intervals where X_i is increasing with rate c_i are of course production times for the product in question, the amount produced in one such interval is called one lot.

No backlog is allowed in the model, and so cumulated production must not fall below cumulated demand at any time t:

$$\forall t \in [0,T], i = A, B : X_i(t) \geq D_i(t). \quad (1)$$

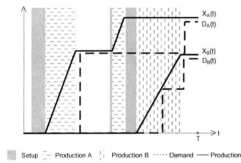

Fig. 2. Demand and production.

Immediately before the production of any lot, appropriate setup has to be performed, which for a changeover from product i to product j takes t_{ij}^S time units and costs k_{ij}^S monetary units. (In the non-sequence-dependent setting of the original model from Jodlbauer (2006a), each setup of e.g. product A took $t_{S,A}$ time units and cost $k_{S,A}$ monetary units.)

Only one product at any time can be dealt with, and for this product, there can be either setup or production activity, but not both.

If some schedule satisfies all of these conditions, it is called *feasible*.

Among all feasible schedules, the aim is to find one which minimizes the cumulated setup and holding costs. Setup costs can be calculated by simply adding them up, and the holding costs are obtained by multiplying the area between X_i and D_i with the holding costs per product unit per time unit for the appropriate product, k_i^I:

$$cost = S + H$$

$$S = \sum_{\substack{i,j \text{ changeovers} \\ \text{from } i \text{ to } j}} k_{ij}^S \qquad (2)$$

$$H = \sum_i k_i^I \int_0^T X_i(t) - D_i(t)\, dt$$

4 DEMAND MODIFICATION

In Figure 2, it can be clearly seen that the increase rate of D_i is at times far greater than the production capacity c_i: In fact, the increase at the due dates is instantaneous, which would amount to an increase rate of infinity. Such points, and the end points of intervals where the demand rate is strictly greater than the capacity, are called *Demand Entry Points* (for exact definitions, please refer to Jodlbauer 2006a as well as Jodlbauer and Weidenhiller 2006).

If there are no Demand Entry Points, all backlogs can be detected by checking at lot start times only. Otherwise, the Demand Entry Points have to be checked as well, which brings along all kinds of difficulties, especially for the solution algorithms (see Figure 3).

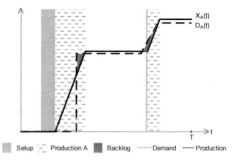

Fig. 3. In the presence of Demand Entry Points (as within the left-hand lot), it is not sufficient to check for backlogs at the lot start times only.

Jodlbauer and Weidenhiller (2006) prove that these problems can be avoided by modifying the shape of the cumulated demand, and that the optimal schedule with respect to this modified demand $\tilde{D}_i$ is also the optimal schedule with respect to the original demand D_i.

The way to achieve this is very simple, using cumulated demand: Starting from each Demand Entry Point, draw a line with slope c_i to the left until this line first intersects with the graph of the original demand. In the interval between this intersection and the Demand Entry Point, the modified demand $\tilde{D}_i$ is described by this line; outside all such intervals, the modified demand is equal to the original demand. For an illustration, refer to Figure 4.

For any given schedule, the area between X_i and $\tilde{D}_i$ is smaller than the area between X_i and D_i, so that the costs calculated using (2) are lower than the actual costs. The difference however is a constant, independent of the schedule in question.

If, with either the old or the new demand, for two schedules A and B the following holds true, this is automatically also true for the other demand:
- A and B are feasible
- A is cheaper than B.

Therefore we can conclude that optimality of a schedule is not influenced by a changeover to the new modified demand. Details of the proofs can be found in Jodlbauer and Weidenhiller (2006).

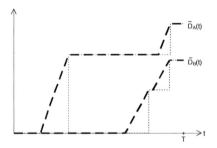

Fig. 4. Demand without Demand Entry Points.

5 NEAR-OPTIMAL SCHEDULE BY APPLYING A SIMPLE RULE

5.1 Need for a Start Solver

The main scheduling algorithm we are developing takes the form of a specialized improvement heuristic, as presented at last year's conference (Reitner and Weidenhiller 2005). This algorithm therefore requires a starting point, for the generation of which some useful methods have been developed.

5.2 EPL approach

One approach, also presented last year, uses as a basis the EPL Formula by Taft (1918), which was developed for one product with constant demand. While this approach performs sufficiently well, there is no reason, especially for discrete demand, why the resulting schedule should be close to optimal.

5.3 Holding Cost and Production Capacity (Discrete Demand)

In the case of discrete demand, considering a simplified problem without setups (i.e. setup times and setup costs are zero) allows one to immediately generate the optimal schedule: The product with the highest value $k_i^I c_i$ (holding costs multiplied by capacity) always must be produced as close to the corresponding due dates as possible.

Having inserted all orders for this product into the schedule, the product with the next highest value of $k_i^I c_i$ should be dealt with in the same manner, never changing any order of the first product. This is to be continued down to the product with the lowest value of $k_i^I c_i$, which is scheduled last.

Optimality of this schedule is proven by showing that a schedule is necessarily suboptimal if at any time a product is produced while another product with a strictly greater value of $k_i^I c_i$ is still in stock.

For problems where setups are nonzero but still comparatively small in respect to production times, this method will often still turn out the optimal schedule, and in any case can be expected to produce a solution close to the optimum.

Computational tests confirm this behavior; however, for problems with large setups and small differences between the values $k_i^I c_i$, the method spreads production over markedly too many lots, resulting in schedules clearly not optimal. In these cases, further optimization, e.g. using the improvement algorithm introduced by Jodlbauer (2006b), is necessary.

6 COMPUTATIONAL RESULTS

6.1 Comparison with Previous Performance

At last year's conference the results of a benchmark with APO from SAP were presented (Reitner and Weidenhiller 2005). Using the same data, the new start algorithm outlined in Section 5.3 was tested. The results (see Table 1) clearly point out the usefulness of the new method – the new algorithm is very much faster and instantly produces markedly cheaper schedules (in the table this is expressed by a higher percentage value of improvement over the solution obtained using APO).

Table 1. Comparison with Previous Performance.

Number of products	Previous best Calc. time[†]	Cost[‡]	New method Calc. time[†]	Cost[‡]
3	13,34	8,7 %	<0,1	11,6 %
4	17,55	2,3 %	<0,1	11,0 %
5	47,14	9,2 %	<0,1	13,6 %

[†] In seconds.

[‡] Cost is given in percentage of improvement over the solutions obtained using APO.

6.2 Scientific Benchmark

The second group of tests are taken from Jordan and Drexl (1998), where they compare their own branch & bound approach with several other sources, thus also providing a small "library" of benchmark problems. Furthermore, we were able to obtain their solver code, which allowed us to compare performance in similar conditions. The models however differ in several important respects, that is:
- Time-continuous versus time-discrete, and
- whether idle time between setups is punished.

Jordan and Drexl calculate in a time-discrete manner, whereas the model presented here is time-continuous. Holding costs calculated by the former are therefore always smaller than when calculated by the latter.

While both models consider sequence-dependent setups, Jordan and Drexl make a further distinction in case there is idle time immediately before the production of some lot. Setup then is not calculated as a changeover from the previous product, on the contrary it is considered as a setup from scratch, which is often more expensive. In this way, frequent occurrence of even small amounts of idle time is punished, often even severely.

The method of comparison we chose is therefore as follows: Both solvers calculate their optimal solution according to their own objectives, but the comparison uses the objective function (2) for both models.

Although Jordan and Drexl base their test problems on Cattrysse et al (1993), Fleischmann (1994) and Salomon et al (1997), a great part of the test instances do in fact differ from the original problems:

The data for the comparison with Fleischmann (1994) were randomly generated by Jordan and Drexl themselves and then solved both with the solver provided by Fleischmann and with their own code.

The problems which Jordan and Drexl obtained from Salomon et al (1997) were modified concerning the setup times and the time horizon: The setup times were changed so as to satisfy the triangle inequality, that is, that a changeover from product A to product C and then to product B always requires a total time at least as great as a changeover from A to B directly. The time horizon was then adapted to maintain the original capacity utilization (see Jordan and Drexl 1998).

Tables 2, 3 and 4 present an overview of our test results, Table 5 contains the sums of the previous three tables. In 79 % of the problems studied, both the solver of Jordan and Drexl (JD) and our own code (Jod) come up with feasible solutions; in 49 % Jod finds the better one. In 7 % of the cases, however, Jod did not find a feasible solution while JD did; this can at least partially be explained by the fact that during the search Jod allows schedules which start before zero, which is the beginning of the time horizon. Infeasibility is always caused exactly by this. A method to deal with this problem more efficiently should be developed, but is probably not as straightforward as it may seem. The cases where Jod found a feasible solution, while JD did not, can be explained by the different treatment of setups after an idle period.

The difference in computation time is still rather large: While JD requires an average of 3 s per problem, Jod requires 770 s on average. Several points probably contribute to this difference:

- While JD exclusively deals with discrete data, Jod is conceived for a greater range of problems,
- The structure of the combinatorial search is still very much in development and not yet to be considered either final or entirely satisfactory,
- It should be possible to further optimize the parameters of this search.

Table 2. Results of Benchmark with Problems from Cattrysse et al (1993).

Feasibility JD	Jod	Who is better	Number of Problems	Percentage of Problems	Avg. diff. in objective
N	Y	-	9	3 %	
Y	N	-	11	4 %	
N	N	-	45	17 %	
Y	Y	Jod	152	57 %	11 %
Y	Y	JD/Jod	2	1 %	0 %
Y	Y	JD	51	19 %	-6 %
			270	100 %	

JD: Solver of Jordan and Drexl (1998)

Jod: Improved solver of Jodlbauer (2006b)

JD/Jod: Both JD and Jod evaluate to the same objective value

Percentage of problems values are rounded, which accounts for the slight discrepancy in the total.

Differences in objective calculated with solution from JD as base, positive value indicates that Jod produced the better solution.

Table 3. Results of Benchmark with Problems from Fleischmann (1994).

Feasibility JD	Jod	Who is better	Number of Problems	Percentage of Problems	Avg. diff. in objective
Y	Y	Jod	56	47 %	3 %
Y	Y	JD/Jod	6	5 %	0 %
Y	Y	JD	64	48 %	-12 %
			120	100 %	

Notation as in Table 2.

Table 4. Results of Benchmark with Problems from Salomon et al. (1997).

Feasibility JD	Jod	Who is better	Number of Problems	Percentage of Problems	Avg. diff. in objective
N	Y	-	1	1 %	
Y	N	-	24	20 %	
N	N	-	18	15 %	
Y	Y	Jod	40	33 %	5 %
Y	Y	JD/Jod	0	0 %	-
Y	Y	JD	38	31 %	-8 %
			121	100 %	

Notation as in Table 2.

Table 5. Results of all Benchmarks with Jordan and Drexl (1998) summed up.

Feasibility JD	Jod	Who is better	Number of Problems	Percentage of Problems	Avg. diff. in objective
N	Y	-	10	2 %	
Y	N	-	35	7 %	
N	N	-	63	12 %	
Y	Y	Jod	248	49 %	8 %
Y	Y	JD/Jod	8	2 %	0 %
Y	Y	JD	147	29 %	-9 %
			121	100 %	

Notation as in Table 2.

7 SUMMARY AND OUTLOOK

In this article, several improvements to the multi-item, single-machine dynamic scheduling problem and appropriate solution algorithms have been presented. Expected benefits are a more concise and more easily understandable model, better quality solutions and a reduction in required computation time.

Benchmarks partially confirmed these expectations and also pointed the way to further improvements and research.

In consequence, further research should be directed towards optimizing the combinatorial search and further improving computational speed. In addition, more benchmarks should be performed. It would be especially desirable to be able to benchmark against problems with continuous demand, of which as yet we have been unable to find anything in the literature.

Research should also focus on the further investigation in and development of the new start solver introduced in Section 5.3.

8 LIST OF SYMBOLS USED

Table 6. Table of Symbols.

Symbol	Description
$D_i(t)$	cumulated demand of product i at time t
$X_i(t)$	cumulated production of product i at time t
T	end of the planning horizon
c_i	capacity for production of product i
k_i^I	unit holding cost of product i
$k_{i,j}^S$	setup cost (changeover from product i to j)
$t_{i,j}^S$	necessary setup time for changeover from product i to j
S	aggregated setup costs
H	aggregated holding costs

9 REFERENCES

[1] Bomberger, E. (1966) A dynamic programming approach to a lot size scheduling problem. *Management Science* 12, 778-784.

[2] Cattrysse, D., Salomon, M., Kuik, R. and van Wassenhove, L.N. (1993) A dual ascent and column generation heuristic for the discrete lotsizing and scheduling problem with setup-times. *Management Science* 39, 477-486.

[3] Fleischmann, B. (1994) The discrete lot-sizing and scheduling problem with sequence-dependent setup-costs. *European Journal of Operational Research* 75, 395-404.

[4] Hanssmann, F. (1962) *Operations Research in Production and Inventory Control*. New York, John Wiley.

[5] Harris, F. W. (1913) How many parts to Make at once. *Factory: The Magazine of Management* 10(2), 135-136, Reprint, *Operations Research* 38(6), 1990, 947-50.

[6] Holt, C.C., et al (1960) *Planning Production, Inventories and Workforce*. Prentice-Hall, Engewood Cliffs.

[7] Jodlbauer, H. (2006a) An approach for integrated scheduling and lot-sizing. *European Journal of Operational Research*, 172/2, 386-400.

[8] Jodlbauer, H. (2006b) *An efficient algorithm for the integrated multi-item dynamic scheduling problem*. Working paper.

[9] Jodlbauer, H. and Weidenhiller, A. (2006) *Transforming demand without changing optimality*. Working paper.

[10] Jordan, C. and Drexl, A. (1998) Discrete Lotsizing and Scheduling by Batch Sequencing. *Management Science* 44, 5 (May. 1998), 698-713.

[11] Kimemia, J.G. and Gershwin, S.B. (1983) An algorithm for the computer control of a flexible manufacturing system. *IIE Transaction* 15(4), 353-362.

[12] Kogan, K. and Perkins, J.R. (2003) Infinite horizon production planning with periodic demand: solvable cases and a general numerical approach. *IIE Transaction* 35, 61-71.

[13] Reitner, S. and Weidenhiller, A. (2005) Testing a new approach for the multi-item dynamic scheduling problem. *Proceedings FH Science Day*, Steyr, Shaker 2005, 131-137.

[14] Salomon, M. (1991) Deterministic lot-sizing models for production planning. *Lecture Notes in Economics and Mathematical Systems*, Springer-Verlag.

[15] Salomon, M., et al. (1997) Solving the discrete lotsizing and scheduling problem with sequence dependent set-up costs and set-up times using the Travelling Salesman Problem with time windows. *European Journal of Operational Research* 100, 3 (August 1997), 494-513.

[16] Taft, E.W. (1918) Formulas for Exact and Approximate Evaluation – Handling Cost of Jigs and Interest Charges of Product Manufactured Included. *The Iron Age* 1001, 1410-1412.

[17] Wagner, H.M. and Whitin, T.M. (1958) Dynamic Version of the Economic Lot Size Model. *Management Science* **5**(1), 89-96.

[18] Wiendahl, H.-P. and Springer, G. (1988) Manufacturing process control on the basis of the input/output chart - a chance to improve the operational behaviour of automatic production systems. *International Journal of Advanced Manufacturing Technology*, **3**(1), 55-69.

[19] Wiendahl, H.-P. (1997) *Fertigungsregelung: logistische Beherrschung von Fertigungsabläufen auf Basis des Trichtermodells*. München, Wien: Hanser.

Automated Generation of Simulation Models from ERP Data – An Example

Gabriel Kronberger[a,]*, Andreas Weidenhiller[b]

[a] Upper Austria University of Applied Sciences, Hauptstraße 117, A-4232 Hagenberg, AUSTRIA,
[b] Upper Austria University of Applied Sciences, Wehrgrabengasse 1-3, A-4400 Steyr, AUSTRIA

ABSTRACT

Motivation: Simulation is an important modeling method for the design and analysis of production systems. Building simulation models is a complex manual process typically performed by simulation experts. Automating parts of this process could lead to more rapid and more precise development of simulation models. Integration of Enterprise Resource Planning (ERP) systems into simulation is a possible way to enable automated generation of simulation models. Integrating the ERP system into simulation is also likely to improve simulation results and makes it easier to apply these results in the actual system.

Results: The authors develop a software system that improves the modeling process in simulation by automating parts of the manual modeling process. The core component of the software is the *Model Generator* which reads preexisting production data from the ERP system and builds a simulation model based on these data.

Availability: The software described in this paper is still in the design and implementation phase. Currently there are no results regarding the capability of the software to build simulation models from actual data of real production system.

Contact: gabriel.kronberger@fh-hagenberg.at

1 INTRODUCTION

Simulation is a valuable technique to analyze production processes in manufacturing. However building accurate simulation models is a delicate process which is typically carried out manually by simulation experts. This process involves three distinct phases [1]:

- Data collection: Building a model requires to collect and analyze data from the real system. The quality of data acquired in this step is an important factor for the quality of the simulation model. Of course other factors in the modeling process also have an influence on the quality of the simulation model. High quality data is no guarantee for an accurate simulation model.

- Design and Implementation: The design of the model depends on the requirements for the simulation and the level of abstraction that is needed. For successful simulation projects it is very important to have a clear definition of the metrics that have to be provided by the simulation.

- Verification and Validation: Design and implementation of the model have to be verified and validated to make sure that results of the simulation coincide with the actual system. The verification phase and the implementation phase are strongly interconnected. Changes in the implementation have to be

---

*to whom correspondence should be addressed

verified and verification provides hints to improve the implementation.

In average simulation projects a lot of effort is put into the data collection phase. However, ERP systems already contain most of the data required for the model. Therefore it is natural to retrieve the necessary data directly from the ERP system. The feasibility of this idea has been researched in [2], [3] and [4].

Automating the data collection is an important step to reduce the time spent in modeling. However, automating the design and implementation phase as well and combining it with automated data collection would minimize the effort needed in modeling. This allows the user to concentrate on verification and validation which is very difficult or impossible to do automatically.

For automating the design and implementation phase it is necessary to provide frequently used atomic building blocks in simulation and a component that composes these building blocks together based on data collected from the ERP system.

Building a model from such atomic building blocks also has the benefit that validation of such models becomes easier, because it is sufficient to thoroughly review and validate the commonly used building blocks only once. In simulation models that reuse those building blocks only the variable high level structure has to be reviewed and validated again. Modern simulation software often includes such building blocks (e.g., stations, transporters, queues and conveyors) but the users have to assemble concrete models from such building blocks manually. Typically simulation is run externally without connection to any part of the production system. However, ERP systems are a central controlling component in production systems and have a significant influence on the processes in the system. Therefore a successful simulation model must also either simulate the ERP system or integrate the ERP system into the simulation process. Integration is the way to go to achieve highly accurate simulation results which can be directly applied to the operative production system.

1.1 Related Work

Automated generation of simulation models can simplify the interactive analysis of manufacturing scenarios [5]. [4] discusses the possible benefits and problems of automated data collection from ERP systems. Automated simulation model generation is also a topic in real-time simulation-based shop floor control [6, 7].

This paper gives an design overview of the software and its components in section 2. Section 3 describes which data the generator uses to construct an example model and how it accesses these data. Sections 4 and 5 show the possibilities and problems of automated simulation model generation and give an outlook on possible future research topics.

2 OVERVIEW OF THE PROPOSED SOFTWARE

An overview of the components of the system is shown in figure 1. The core component is

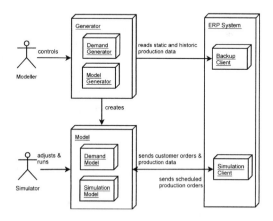

Fig. 1. Overview of the components of the software system.

the *Generator* which includes two separate components a *Demand Generator* and a *Model Generator* that also includes a set of predefined building blocks for models. The *Model Generator* reads historic production data from the *ERP System* and creates a *Simulation Model* of the production system. The *Demand Generator* component reads historic sales data from the *ERP System* and creates a *Demand Model* of the external demand for products.

2.1 ERP System

The *ERP System* has two separate clients which hold different data: a *Backup Client* and a *Simulation Client*. This separation is necessary because a simulation run changes the data in the *ERP System*. By separating the clients for the *Generator* and the *Simulation Model* it is possible to generate multiple models based on the same data. The *Generator* reads data from the *Backup Client* and the *Model* uses the separate *Simulation Client* to read and write data. The *Simulation Client* is reset between each individual simulation runs.

2.2 Model Generator

The *Model Generator* connects to the *Backup Client* to read static production data. Based on this data the *Model Generator* uses the building blocks to create the structure of a *Simulation Model*. In a second step the *Model Generator* reads historic production orders and production order confirmations from the *Backup Client*. Based on the historic production orders and production order confirmations the *Model Generator* estimates stochastic parameters of the simulation model (e.g., processing time for operations, and the failure rate and failure duration of resources).

An important feature of the *Model Generator* is that it is able to build *Simulation Models* that use different production control systems. The user can choose whether the simulation model uses MRP, Kanban or CONWIP control.

2.3 Demand Generator

The *Demand Generator* reads historic customer orders from the *Backup Client*. Based on a statistical analysis of the historic customer orders the *Demand Generator* sets the parameters of the *Demand Model* which is used for the simulation of demands.

2.4 Demand Model

The *Demand Model* has a list of products and for each product it keeps a list of time intervals with different demand parameters. The demand for a product is determined by three stochastic parameters which are estimated by the *Demand Generator* from historic sales data and subsequently adjusted by the user if necessary. These parameters are:

1. Amount (quantity) of the product in each customer order.
2. Rate at which new customer orders arrive in the system.
3. Delivery time is the time span between arrival of a new customer order and its due date.

Based on these stochastic parameters the *Demand Model* generates a set of new customer orders with due dates for each product each day. The customer orders are sent to the *Simulation Client* of the *ERP System* which generates and schedules production orders based on the customer orders.

2.5 Simulation Model

The discrete-event based *Simulation Model* can be used to simulate the production process for a given time span. The *Simulation Model* does not include scheduling or control algorithms. Instead it can integrate an *ERP system* into simulation. The *ERP System* schedules production orders which are executed by the *Simulation Model*. Before starting a simulation run the user can change specific parameters of the *Simulation Model*. It is possible to adjust probability distribution parameters for stochastic events (e.g., machine failures, processing times, setup times...). It is also possible to choose different dispatching rules for each station.

More details about the implementation of the simulation model are given in [8].

2.6 Limitation of the simulation model

As in every simulation model a few details are not considered because the model would become too complex otherwise.

The software only supports manufacturing scenarios where discrete material items are processed. Currently there is no support for continuous material flows (e.g., chemical processing) or continuous demands. The software also models transport processes, however material transports are just special cases of normal material processing steps. The software does not support automatic transport vehicles, path finding or collision detection. Employees are modeled as a pool of human resources. They cannot walk around in the simulated production space. Employees who are available at a plant at a given time can move to any given workstation instantaneously.

2.7 Implementation

The software is implemented in the Java^{TM} programming language. It uses the SAP^{TM} Java Connector to communicate with SAP R/3^{TM} systems. The discrete event-based simulation models use the core framework and the enterprise library of the Java-based simulation software AnyLogic^{TM}.

More details on the implementation of the software and the generic simulation model including its basic building blocks are given in [8].

3 DATA COLLECTION FROM THE ERP SYSTEM

SAP^{TM} systems provide so called *Business Objects* for data that is often used externally. External applications can access these business objects through remote function calls. The model generator also uses business objects when possible. However not all data that are needed are accessible through business objects. The model generator reads these data directly from the corresponding database tables. The generator uses the following data structures of the SAP^{TM} server to access production data:

- Plants: table `T001W` (Plants/Branches)
- Factory calendars: tables `TFACS` (Factory Calendar) and `TFACD` (Factory Calendar Definition)
- Setup-Matrix: table `TCYTM`
- Employee data: Business Objects `Employee`, `EmployeeAbsence`, `TimeAvailSchedule` (Employee Availability Schedule), `EmployeeAttAbs` (Employee attendance or absence), `Qualification` (Person's qualification), `Qualificationtype`
- Stations: table `CRHD` (Work Center Header)
- Storage locations: table `T001L` (Storage Locations)
- Production orders: Business Objects `ProductionOrder` and `ProdOrdConfirmation` (Production order confirmation). The model generator uses these data to estimate the actual processing- and setup times and the rate and duration of machine failures. Production order confirmations also contain data about scrap material.
- Customer orders: Business Object `SalesOrder`. The demand generator uses the fields date, product, volume and delivery date to create demand models.

A simplified example company is used to test the functionality of the model generator and the simulation model. The ERP system holds data of this company which sells two different products, a large and a small shelf. Demand for the two final products is higher in summer than in winter. Work is done in one eight-hour shift each day for five days a week, no work on national holidays.

The company has one plant with five workstations where material is processed in a sequence (flow shop layout). Each station has a different scrap rate, failure rate and repair duration. A setup-matrix defines the setup time between different operations for each workstation.

Five employees work at the plant, each of them can work at two or three different workstations. Employees have different absence rates and absence durations.

At some workstations material is processed by a machine which needs an operator (e.g., drilling) while at other stations a worker processes the material and optionally uses tools in this process (e.g., finishing).

Material is transported on pallets that are stored in buffers between stations. Buffers have a limited pallet capacity. If an output buffer is full the workstation cannot start to process new material.

4 DISCUSSION

The close connection of the model generator and the simulation model to the ERP system could also be useful to tune ERP system parameters. Typically an ERP systems is adjusted to a specific production system by a "customizing expert". However, such experts sometimes set parameters which are not optimal for the specific production system because they lack expertise in production management which is necessary to optimally tune ERP system settings. It is also the case that ERP system settings are not changed after the initial implementation phase even when changes in the production system occur.

The automated approach described in this paper heavily depends on the consistency of data in the ERP system. If data is missing, inaccurate or inconsistent it is not possible to generate an accurate model. This is also an issue for the estimation of probability distribution parameters. If not enough historic data are available to estimate distribution parameters with enough significance the user has to set the parameters manually.

Another issue of the system is the relatively long runtime of simulation. Trying various changes to the production system in the simulation is not practical if each run lasts more than a few hours. Runtime can be reduced by limiting the simulation interval to a few weeks or by removing complexity from the system by aggregating stations to station groups and products to product groups.

5 FUTURE WORK

Currently the authors are working on the implementation of the simulation model generator and the generic simulation template. The next steps are the automated generation of simulation models directly from ERP data and validation of the generated models by comparing results from simulation with the actual processes in the production system.

When the system has reached a corresponding level of sophistication the authors plan to use the simulator for the further refinement of a dynamic lot-size optimizing scheduler [9]. The idea is to benchmark this scheduler and other schedulers in simulated production scenarios. Recently a benchmark of this scheduler and the SAP APO^{TM} component showed that the former achieves better results [10]. However, it is difficult to draw objective conclusions because of differences in the benchmarked schedulers. The combination of a realistic simulation model with the definition of a clear interface between simulation and scheduling could lead to more objective comparisons of the performance of different schedulers. In the future it will also be interesting to integrate simulation into the scheduling algorithm in order to improve the scheduler results.

With the current design of the software the generated models are rather static. The user can adjust some parameters after model generation but it is not possible to change the whole structure of the model for instance to add additional stations or buffers. A possibility to overcome this limitation is to directly output model files that can be read by simulation software that provides a graphical environment for modeling. This would give the user full control over the generated simulation models and it would be possible to change and extend a basic model that was automatically generated by the proposed model generator. AnyLogic™ a simulation software with a graphical modeling environment could be a possible target for this because it uses an open XML file format to store its simulation models. Generating a XML document that is compatible to the AnyLogic™ file format seems possible.

Another interesting research topic is the correct aggregation of stations to station groups and products to product groups by the model generator. Aggregation has two potential benefits: It allows to speed up the simulation of real manufacturing scenarios and makes it possible to create simulation models with different levels of abstraction. However the correct aggregation of stations is not trivial because the behavior of a whole station group is radically different from the clear-cut behavior of a single station.

6 ACKNOWLEDGEMENTS

The work described in this paper was done within the project "Integrated Scheduling" funded by the provincial government of Upper Austria.

7 REFERENCES

[1] Law, A. M. and Kelton, D. W. (2000) *Simulation Modelling and Analysis*. McGraw-Hill Education - Europe.

[2] Perera, T. and Liyanage, K. (2000) Methodology for rapid identification and collection of input data in the simulation of manufacturing systems. *Simulation Practice and Theory*, **7**, 645–656.

[3] Robertson, N. H. and Perera, T. (2001) General manufacturing applications: feasibility for automatic data collection. *WSC '01: Proceedings of the 33nd conference on Winter simulation*, Washington, DC, USA, pp. 984–990, IEEE Computer Society.

[4] Robertson, N. and Perera, T. (2002) Automated data collection for simulation? *Simulation Practice and Theory*, **9**, 349–364.

[5] Kulvatunyou, B. and Wysk, R. A. (2001) Computer-aided manufacturing simulation (CAMS) generation for interactive analysis-concepts, techniques and issues. *Proceedings of the 2001 Winter Simulation Conference*, pp. 986–976.

[6] Son, Y. J. and Wysk, R. A. (2001) Automatic simulation model generation for simulation-based, real-time shop floor control. *Computers in Industry*, **45**, 291–308.

[7] Son, Y. J., Joshi, S. B., Wysk, R. A., and Smith, J. S. (2002) Simulation-based shop floor control. *Journal of Manufacturing Systems*, **21**, 380–394.

[8] Kronberger, G., Weidenhiller, A., Kerschbaumer, B., and Jodlbauer, H. (2006) Automated simulation model generation for scheduler-benchmarking in manufacturing. *Proceedings of the 2nd European Modeling and Simulation Symposium (EMSS06)*.

[9] Jodlbauer, H., Reitner, S., and Weidenhiller, A. (2006) Reformulation and solution approaches for an integrated scheduling model. *ICCSA 2006*, vol. 3984 of *LNCS*, pp. 88–97, Springer.

[10] Reitner, S. and Weidenhiller, A. (2005) Testing a new approach for the multi-item dynamic scheduling problem. *Proceedings FH Science Day*, pp. 131–137, FH Steyr, Shaker.

Java and all Java-based trademarks and logos are trademarks or registered trademarks of Sun Microsystems, Inc., in the U.S. and other countries.
SAP, ABAP and BAPI are the trademarks or registered trademarks of SAP AG in Germany and in several other countries.
AnyLogic is a registered trademark of XJ Technologies Company.
All other trademarks are property of their respective owners.

Total Effective Equipment Productivity - The Key to Maximizing the Efficiency of the Industrial Production

Martin Schöffer[a*], Christian Weger[b]

[a] Upper Austria University of Applied Sciences, Wehrgrabengasse 1-3, A 4400-Steyr, AUSTRIA
[b] Upper Austria University of Applied Sciences, Wehrgrabengasse 1-3, A 4400-Steyr, AUSTRIA

ABSTRACT

Motivation: Especially in manufacturing companies which own capital-intensive manufacturing facilities in high-wage-countries like Austria or Germany maximum efficiency of equipment is essential. Assuming that the costs for raw material and equipment are independent of geographic location, the disadvantage of high personnel costs can be compensated for only by excellent logistic features such as exact delivery dates, short delivery times, high quality and in particular by a high degree of utilization of the machines. As a result of the permanent cost pressure in the industrial sector of automotive-part-suppliers and especially in the die casting industry, optimal utilization of the manufacturing equipment constitutes a critical competitive advantage.

The key metric, Total Effective Equipment Productivity (TEEP), shows the ratio of equipment runtime for conforming goods in relation to the calendar time and is very useful to describe a company's degree of equipment utilization. A detailed analysis of the different losses helps to identify potential for increasing capacities and therefore for increasing turnover and profit.

An investigation of scientific papers showed a wide variety of – more or less - different definitions and types of time-losses and very little data on actual TEEPs or other key metrics of equipment efficiency like OEE (Overall Equipment Efficiency, shows ratio of runtime to scheduled time). Further investigation showed that it is not possible to compare OEE´s of different companies because the bases of the OEE differ widely. As a result of the fact that the basis of TEEP is calendar time, a comparison of different manufacturing systems is possible if the different types of losses are categorised in a clear procedure.

So a TEEP benchmark based on data reports of approx. 100 die casting cells in 7 participating companies from Austria and Germany has been developed.

The different types of losses have been identified and categorized. The TEEP of each die casting cell, an average for each company and the average of all companies have been calculated. Based on these benchmark figures it was easy to determine the Best in Class and the range of utilization typical for the die casting industry

After the difficult process of collecting and preparing the data, the analysis of the results delivers even more interesting coherences between different factories which are worth looking at in more detail in the future.

Results: A typical, medium-sized fabrication cell consists of furnace, die casting machine, deburring press, cooling device and a handling robot and demands investment costs of approx. € 1,400,000 and is operated at a cost rate of approx. € 100 – 120 per hour while the ratio of direct labour costs is below 10 - 15%.

The survey of the TEEP in die casting companies in Austria and Germany showed that a lot of the production systems are operated with average TEEPs just below 50% with a range from 14% to 70%. The Best in Class company with 70% TEEP gains more than 6,100 hours runtime for each of its die casting cells while the industry average is 48% or only 4,200 hours. The causes of different types of downtime (in descending order) are: lack of operators, troubles in operations management (scheduling, tooling, rough material), technical downtime, planned downtime (cleaning, maintenance), change of product (planned or unplanned) and waste (scrap, rework, short downtimes, not ideal cycle time).

Detailed cause – effect analysis of these different causes of losses showed opportunities to reduce the downtimes by at least 25 – 30%.

The results show that there is a lot of room for improvement and the ranking list of productivity consuming problems.

Availability: The study of the TEEP in the die casting sector will be available as an anonymous report at the beginning of 2007.

Contact: martin.schoeffer@fh-steyr.at

1 INTRODUCTION

The original purpose of the key figures OEE and TEEP was to evaluate the progress of the methods of Total Productive Maintenance (TPM) by measuring the effectiveness of individual equipment, entire processes or whole factories.

Both key figures show the combined effects of losses due to a poor quality level, reduced production speed rate, downtime caused by technical malfunction, problems of operations or production quality and stop time.

The TEEP (total effective equipment productivity) describes the runtime of a manufacturing system in relation to calendar time i.e. 24 hours per day and 365 days per year. The actual productive runtime, also called output, is equal to the equipment time needed to produce conforming goods for the first time in the ideal cycle time. This means, that the TEEP shows the percentage of equipments runtime which is used to gain money.

The Overall Equipment Effectiveness (OEE) is a similar key metric to describe the degree of utilization of production equipment but on the basis of scheduled time alone.

OEE therefore ignores planned downtimes like shut-downs or maintenance stops, meetings and training of production staff or experiments, development of new production methods and online rework and differs very strongly from company to company.

As TEEP categorizes all events around the clock it shows unexploited capacities and the possibilities to minimize the hidden factory. Because of the clear basis of the calendar time there are no problems in the definition of planned or unplanned downtimes and therefore the TEEP of different companies, production sites or systems in the same sector of trade can be compared easily.

Next to the opportunity to bench mark a company's manufacturing performance by TEEP the analysis of the different types of losses and especially their causes is the most important application of this tool.

2 CALCULATION OF OEE AND TEEP

Nakajima defines OEE within the TPM philosophy as a metric or measure for the evaluation of equipment effectiveness [Nakajima, 1988]. This means that OOE and furthermore TEEP tries to identify production losses and other direct and "hidden" costs.

The simplest method to calculate the OEE is

OEE [%] = number of conforming goods x ideal cycle time / scheduled time

Example (A): scheduled time: 480min; ideal cycle time: 0,3 min; good units: 1125
OEE = (1125 x 0,3) / 480 = approx. 70%

Therefore detailed records of events and time are not necessary to calculate the OEE in the beginning. To use it as an effective management tool it is necessary to know the three main components of OEE:

(1) Availability – The ratio of machine run time to scheduled production time.
(2) Speed – The ratio of actual running speed to the machine's maximum speed.
(3) Yield – The ratio of good material produced to the total material used.

Finally the knowledge of availability requires a data collection of all kinds of losses of runtime such as downtime and stops. Especially the analysis of these data and their causes will show profitable opportunities to improve the efficiency of equipment.

The combined effect of the three components is expressed as

OEE [%] = Availability [%] * Speed [%] * Yield [%]

("Speed" is often also named "Performance" and "Yield" also as "Quality Rate").

Example (B):
A press operates at 70 strokes per minute while the designed cycle speed is 100 strokes per minute: Speed / Performance Rate 70%.
90 good stampings out of every 100 Yield / Quality Rate 90%.
Scheduled production time 8 hours, technical downtime one hour, Availability 87, 5%.

OEE [%] = 0,70 x 0,90 x 0,875) = 55%.

Also the calculation of the TEEP could be done in most companies by easily available data

TEEP [%] = runtime/ calendar time = number of conforming goods x ideal cycle time/ calendar time

Example (C): calendar time: 24 h; theoretical cycle time: 0,3 min; good units: 1125
TEEP = (1125 x 0,3) / 24 x 60 = approx. 23%

The demonstrated examples (A) and (C) for calculation of the OEE (approx 70%, basis 8 hours) and TEEP (approx. 23%, basis 24 hours) show the problems in comparing these two figures directly.

3 CLASSIFICATION AND AMOUNT OF LOSSES IN PAPERS

Losses can be defined as activities that require resources but create no value [Bamber, Castka, Sharp and Motara, 2003].

In order to use the TEEP as an indicator for the size and the causes of the hidden factory, it is essential to identify the different types and shares of losses of asset utilization. Therefore detailed production reports on event and time and a clear classification of the different types of losses, which often vary with type of equipment, process, or production method are very important.

According to Nakajima [Nakajima, 1988] or Hansen [Hansen, 2002] the following types of losses can be identified (and have to be reported) in manufacturing industries with discrete processes: No Production scheduled, excluded time (planned time not scheduled for production. shift change, lunch, planned shutdown), Technical down-time (unplanned; equipment failures), operational downtime (unplanned; not following procedures, accident..), quality downtime (unplanned; nonconforming material, process control problem..), stop time induced (unplanned, external cause: lack of material, staff, tooling..) and stop time operational (planned; changeover, machine cleaning).

Hartmann [Hartmann, 2001] defines six types of losses: no production scheduled, excluded time, changeover, downtime, speed loss and waste loss.

Schwarz adds to the six losses logistical speed losses in chained facilities due to lack of synchronization of production cycles.

Ljungberg [Ljungberg, 1998] defines five different types of losses as follows: Breakdowns, setups, idling and minor stoppage, speed losses and quality losses.

A classification by Jeong and Phillips [Jeong and Phillips, 2001] which is styled for usage in capital-intensive industries defines the following losses: unscheduled maintenance, set-up adjustment, non-scheduled time, scheduled maintenance, R&D time, engineering usage time, WIP starvation time, idle time without operator, speed-losses and quality losses.

Suzuki [Suzuki, 1994] defines "Eight Major Plant Losses" as typical and applicable in continuous process industries which usually produce continuously throughout the year: shut down, production adjustment, equipment failure, process failure, normal production loss, abnormal production loss, quality defects and reprocessing.

For the needs of the die casting industry and regarding the existing production report data the following six different types of downtimes have been defined for this paper:
(1) No production scheduled: holidays, weekends,
(2) Planned downtime: cleaning, maintenance, inspection
(3) Set up: tool changeover including set up and start up; planned or unplanned
(4) Technical downtime: electrics, mechanics, hydraulics
(5) Operational downtime: breakdown, lack of material, staff, tooling, unplanned
(6) Scrap and output loss: waste loss, speed loss, short downtimes

The TEEP is calculated according to the "loss – staircase" as shown in figure 1.

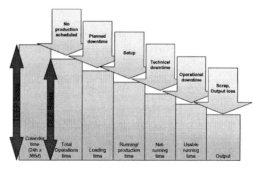

Fig. 1. TEEP loss-staircase.

Few useful figures of OEE values can be found in papers (mostly without published database and calculation method) and practically no data is reported on TEEP.

Hansen summarizes his practical findings of the key figure OEE as following [Hansen, 2002]:
< 65% unacceptable
65 <= OEE < 75 passable
75 <= OEE < 85 pretty good
OEE > 85% world class level for discrete processes
OEE > 90% world class for continuous processes
OEE of 98% is the maximum reported value

Hartmann [Hartmann, 2001] reports improvements of OEE in German tire factories as a result of TPM activities in a range from + 15% to + 43%, the OEE in the beginning is not reported.

Robinson and Ginder [Robinson and Ginder, 1995] report an OEE level for process industries (Japanese chemical plant) rising from 92% (1985) to 99% (1990) following the implementation of TPM.

Wireman [Wireman, 2004] states, that "in most US companies OEE is barely more than 50%" and shows several case studies as
- Ten presses for plastic injection molding: OEE : 24% (!)
- gas compressors OEE 57 to 67%

A study performed by Ljungberg [Ljungberg, 1998] based on 23 machine systems showed an average OEE of 55%, Ahlman [Ahlman,1993] 1993 found an average of 60%, Björklind and Henriksson [Björklind and Henriksson,1994] 1994 showed 73% of 15 paper machines and Wiendahl and Winkelhake [Wiendahl and Winkelhake, 1988] measured in a sample of 140 assembly stations an average OEE of 75%.

4 DATA COLLECTION AND ANALYSIS

Collection of data for reporting should be standardized. This could be done manually or automatically by production data acquisition.

Automatic data collection is expensive, complex and the data is often collected at an aggregated level. A very detailed collection can be achieved with a manual data collection where failures can be carefully examined [Ljungberg, 1998].

For a representative comparison of different companies it is necessary to have the same basis and therefore the TEEP is a proper bench mark. The collection of data must be based on a clear and precise categorization of the types of losses, especially if TEEP is used as a key metric for the manufacturing performance of different companies. The loss-staircase is a very good tool with which to visualize this task.

The next steps are the development of a detailed catalogue of downtime-codes and relating these codes to the appropriate categories of losses and the calculation of TEEP.

The last but most important step should be a detailed cause-effect analysis of the different types of losses to find potential to optimize the utilization of the equipment.

5 RESULTS

A provisional analysis based on data of approximately 50 die casting cells showed the following results of the TEEP:

	Average [%]	Min. [%]	Max. [%]
Calender time	100%	100%	100%
Total Operations time	79%	55%	97%
Loading time	74%	53%	90%
Running/ production time	71%	49%	87%
Net-running time	64%	36%	84%
Usable running time	50%	16%	72%
Output	48%	14%	70%

6 REFERENCES

[1] Bamber, C., Castka, P., Sharp, J., Motara, Y. (2003). Cross-functional team working for overall equipment effectiveness. *Journal of Quality in Maintenance Engineering*, Vol. 9 No. 3, pp.223-239.

[2] Eldridge, S., Garza, J.A., Barber, K.D. (2005). An analysis of OEE performance measurement for an automated manufacturing system. *Proceedings of the 2005 Flexible Automation and Intelligence Manufacturing (FAIM) Conference*.

[3] Hansen, R.C. (2002). Overall equipment effectiveness. *Industrial Press*, New York, pp. 36-45.

[4] Hartmann, E. H. (2001). TPM Effiziente Instandhaltung und Maschinenmanagement. 2. Auflage, *Redline Wirtschaft*, Frankfurt, pp.76.

[5] Jeong, K., Phillips, D. (2001). Operational efficiency and effectiveness measurement. *International Journal of Operations & Production Management*, Vol. 21 No.11, 2001, pp.1404-1416. MCB University Press.

[6] Ljungberg, Õ. (1998). Measurement of overall equipment effectiveness as a basis for TPM activities. *International Journal of Operations & Production Management*, Vol. 18 No.5, pp.495-507. MCB University Press.

[7] Nakajima, S. (1988). An introduction to TPM. *Productivity Press*, Portland, OR.

[8] Robinson, C. J., Ginder, A. P. (1995). Implementing TPM. *Productivity Press*, New York.

[9] Schwarz, J. (2003). Verfügbarkeit verketteter Produktionsanlagen. *Risikominimierung im Anlagenmanagement*. TÜV Verlag, Köln, pp.140-144.

[10] Suzuki, T. (1994). TPM in Process Industries. *Productivity Press*, New York pp. 23-33.

[11] Wiendahl, H.-P., Winkelhake, U. (1988). Permanent automatic supervision of assembly lines. *Developments in Assembly Automation,* March, pp. 289-300.
[12] Wireman, T. (2004). Total productive Maintenance. *Industrial Press,* New York, pp. 46-54.

Workshop:

Secure and Embedded Systems

Chairman: Jürgen Ecker
Thomas Müller-Wipperfürth

Scientific Board:
Michael Bogner, FH OÖ Campus Hagenberg
Hans-Georg Brachtendorf, FH OÖ Campus Hagenberg
Jürgen Ecker, FH OÖ Campus Hagenberg
Johannes Edler, FH OÖ Campus Hagenberg
Eckehard Hermann, FH OÖ Campus Hagenberg
Robert Kolmhofer, FH OÖ Campus Hagenberg
Josef Langer, FH OÖ Campus Hagenberg
Thomas Müller-Wipperfürth, FH OÖ Campus Hagenberg
Markus Pfaff, FH OÖ Campus Hagenberg
Ingrid Schaumüller-Bichl, FH OÖ Campus Hagenberg

NetScanAssistant - Platform Independent Network Scanner as Plug-In in Eclipse

Oliver Hable, Barbara Hauer

GRZ IT Center Linz GmbH, Goethestraße 80, A-4021 Linz

ABSTRACT

For the security and quality of services of a network it is important that an inspection of a network supplies the necessary information as fast as possible. NetScanAssistant is a network scanner platform, which uses existing commercial or free available scanners like Nmap [1] and Nessus [2] to gather information about devices in a network. Nmap is used to find all devices connected to the network and to detect their operating systems. Nessus (for example) executes a system penetration scan on the connected devices to verify if they conform to the security policy of the company. This includes an observation of the patch level, possible targets of attacks, and security issues. NetScanAssistant handles all gathered information and stores it in a relational database. Like the control of the scan, the report management takes place in a perspective of an Eclipse plug-in. The user benefits from some predefined standard queries, which simplify the interpretation of the scan results. The queries give the user a look at the stored data and allow him to search for results and special information. The result of some queries can be stored as an HTML report. The advantage of this is the possibility to create reports for special requirements. These reports can contain all data from a scan or only parts, for example only the results which represent high risk for the network devices. In contrast to other network and security scanners the user is not confronted with a lot of data. NetScanAssistant gives the option to evaluate the results and to search for certain information. That makes the challenge of analysing the information from a network scan more comfortable and flexible. Furthermore, the database allows a long-term storage and comprehensive comparisons, which prove corrections and improvements or other changes of the devices in the network.

Contact: hable@grz.at, hauer@grz.at

1 INTRODUCTION

The security and quality of service of a network is important for many companies. A failure can cause a decrease of productivity. If a company provides several services on their IT infrastructure, their non-availability is unacceptable. That is why safeguarding and steady inspection of a network and its connected devices is a must have. For the inspection of network devices, to prove their compliance to the security policy of the company, a lot of commercial and open source tools like network mappers, network scanners and system penetration scanners are available. There is a wide range of products, but most of them overwhelm the user with a confusing plenty of information. This makes the analysis of the data a time consuming challenge for the security representative or system administrator.

In order to solve this problem and to provide only information which is interesting or necessary for the user, Barbara Hauer developed NetScanAssistant (as her diploma thesis, supported by Oliver Hable, GRZ IT Center Linz GmbH).

2 APPROACH

The focus of a network inspection is to gather as much information as possible about the devices in the network. The approach of the project "NetScanAssistant" is not the creation of a new network scanner with new tests and methods to analyse a system, but of a common platform for existing network and system penetration scanners. An intelligent combination of these available analysing tools makes it possible to use practice-proved techniques to get results with full details that are easy and quick to evaluate through NetScanAssistant. A further advantage of the use of existing and highly reputed scanners are the communities behind them which take care of a steady further development and service. If one of the integrated scanners is updated NetScanAssistant also is up-to-date because it uses always the current version of the scanners.

2.1 Java, a language with many possibilities

Java is an object-oriented programming language developed by Sun Microsystems. It is suited for network applications because of the support of a lot of platforms which makes NetScanAssistant almost platform independent. A wide range of libraries, APIs and frameworks are available to ensure good performance, variable security mechanisms, and that an application runs on different systems without any problems. Existing and well-known Open Source projects like log4j [3], Hibernate [4], JUnit [5] and Eclipse [6] are developed in Java and help to develop applications in Java. NetScanAssistant uses them because they simplify the communication with different systems, integrated network and system penetration scanners, the used relational database, and the user.

2.2 Nessus client API in Java

NetScanAssistant uses a particular Nessus client[1]. The client is Open Source and available as a Java API. The Java client replaces the conventional Nessus client and takes over the functions including the whole communication with the Nessus server, which includes the creation of a key store, the set up of the needed data structure, the duplication of the required files in the client directory, the creation of the Nessus configuration file and the start of a scan. This allows NetScanAssistant as much control as possible over the scan with Nessus and makes it easier to start, stop and continue a scan, which is important for an exact time control. Furthermore, there is the possibility to split the work of scanning up to many Nessus servers for an improved performance. This reduces the time of scanning of large networks where the duration of scan can take a few days.

2.3 Modular design

Design Patterns are about design and interaction of objects and they provide reusable solutions for solving common design problems. To simplify the further development and service of NetScanAssistant, the structure is modular and conforms to very common Design Patterns. This ensures the potential for an application in an object-oriented programming language like Java to acquire the quality for production. Moreover, it helps the compliance of best practices in development to decrease the complexity and to improve the readability of the source code. This increases the quality of the software and supportability of the source code. The modular design makes it simple to expand the functionality of NetScanAssistant. It is conceivable to use not only the existing Eclipse plug-in for the control and analysis of a scan but also new modules to allow this functionality via e-mail, web, PDAs, mobile phones, etc. This involves an upgrade in

---

[1] see Java Nessus Client API.
Url, http://www.sourceforge.net/projects/jnc-api

authentication and user management. To allow these upgrades and other further developments it is important to follow the guidelines and best practices in software engineering. And so this new modules also fit into the modular structure of NetScanAssistant.

2.4 Phases of a scan

The subdivision of a scan in phases enables enhanced scalability and flexibility. Therefore, NetScanAssistant has three different phases:

- **Detection** - The goal is to detect all devices in the network which respond to an ARP query or ICMP packet or where at least one port is listening (open). As a result, all IP addresses are divided into used and unused (free).
- **Operating System Detection** - Detects the operating system running on the found devices.
- **System Penetration** - A system penetration test of the devices helps to verify their compliance to the security policy. The scan discovers the patch level of the system, possible targets for an attack, and security holes.

Every phase can be activated for its own or combined with other phases. This allows an aborted scan to be continued without repeating already finished tests. There are often some sensitive devices in a network. These devices should not be stressed to avoid damage or a shut down of a service. This is why NetScanAssistant provides the option to exclude these devices from a scan by their IP addresses. Supported are IP addresses in version 4. IP version 6 is currently not supported because of low popularity. But an additional implementation is no technical challenge.

The scan of a large network can last a few days. Due to the fact that many devices are not active all the time this results in the necessity of a time control. This means that a scan can be limited to a special date, e.g. from Monday to Friday from 08:00 to 16:00 and no scanning on Saturday and Sunday. The times can be set individually per day to comply with the requirements of different networks and companies. These and all other settings can be defined in a configuration file or via the graphical user interface and are saved in a relational database.

2.5 Detection

In the current version NetScanAssistant is using Nmap [1] for the detection of the devices in a network. Nmap ("Network Mapper") is an open source tool that supports a lot of scan techniques. Real time tests showed that only a few variants of scan techniques and options are qualified for the use in NetScanAssistant. Tests made it obvious that only an ARP ping, an ICMP ping scan or a TCP SYN scan were able to find all devices in the network within an adequate time. The fastest scan was the ICMP ping scan but in some networks this scan did not work because ICMP was deactivated. That is why NetScanAssistant provides this scan technique as an option. To increase the likelihood of success as much as possible and to allow the user to choose more Nmap options NetScanAssistant uses per default the TCP SYN scan with some default ports that are adapted to the network. Per default Nmap also performs an ARP ping before a scan if the user does not explicitly prevents this. For this phase of the scan the user can configure additional Nmap options appropriate to the TCP SYN scan. This allows an exact adaptation of the scan to the target network.

2.6 Operating System Detection

The operating system running on a device is relevant for an overview of the devices in a network or, for example, to compile statistics about the number of devices having a certain operating system. Through comparisons of the results of the system penetration test it is possible to expose false positives and to identify devices which are worth taking a closer look at. A reason why devices can get conspicuous is, for example, when the operating system Windows is detected but the system shows weaknesses of a Linux system. This can happen if the detection of the operating system and software went wrong or the software was ported to the operating system. In both cases it is worth to analyse the device more carefully and exactly.

NetScanAssistant uses Nmap to detect the operating system of a device. One reason for the separation of this scan phase from the system penetration phase, which uses Nessus, is to increase the detection of false positives. Another reason is that Nmap in contrast to Nessus or other tools has provided OS detection for years and it is accordingly reliable and fast.

Exact configurations for a scan and network can also be made in a configuration file or with help of the graphical user interface.

The result of this scan phase is that to each IP address of a found device, the detected operating system is added. If Nmap cannot specify the correct operating system and guesses more than one operating system, then NetScanAssistant saves all of them. If Nmap is unable to guess the operating system of a network device it becomes marked as "unknown". If a device is disabled or disconnected during the scan, an adequate comment is set in the database. This is useful because in the analysis process all devices with a comment like "unknown" or "down" can be filtered and if necessary scanned again to obtain more data about them.

2.7 System Penetration

Obtaining detailed information about a device in the network is the major aim of the system penetration. NetScanAssistant uses the free scanner Nessus [2] for this purpose. Nessus is specialized in system penetration and very fast in this task. It provides detailed reports about possible weaknesses of a device. The decision about the intensity of the tests used and which Nessus plug-ins are used for testing remains up to the user. He has to configure Nessus as usual via a graphical user interface or to choose an already existing configuration.

After the scan NetScanAssistant deletes all files created by Nessus from the system to prevent that other users of the device are able to get any information about the scan. So all data of the scan is saved in the relational database and no sensitive information remains on the system. Among these data is the IP address of the device, port number, port service, port protocol, severity, risk level, a description of the weakness, feasible solutions, references to other information sources, CVE [7] and BID [8] numbers, and the unique identification number of the Nessus plug-in that detected the weakness. NetScanAssistant is not able to filter possible false positives out of the Nessus report because they need to be reviewed individually per device. Therefore the definitive analysis of the scan results is task of the user.

2.8 Data management

All data of NetScanAssistant is saved in a relational database to allow long-term storage. This provides the possibility to determine after months or years what configuration with which options led to which results of a scan of certain target network. In shorter term use it allows to reproduce conclusively what security deficits were found in a network, which devices were affected and if the found weaknesses were remedied until a defined date on which the progress is checked with a new scan. To a certain extent this helps to monitor the compliance to the security policy of the company. Furthermore, the storage in a relational database makes the results available earlier and they can be searched and filtered faster.

Usually the relational database is stored on a server separate from the scan machine. This allows the use of advanced security mechanisms to protect the server with the database from unauthorized access. In this context the integrity of data, access limitations, etc. play an important role and so the level of security has to be adapted to the requirements of the company.

2.9 Report management and result evaluation

The storage of the scan results from NetScanAssistant in a relational database allows an authorized user to get access to data from any system with any tool for database and SQL queries. For more comfort NetScanAssistant provides an Eclipse plug-in with two perspectives. The separation in two perspectives draws a line between the control of a scan and the report management. The perspective for the report management facilitates the analysis of the scan results by predefined queries (templates). This perspective also provides the option to define own SQL or HQL query statements in an extra view. The results of all queries are output to the console and in part they also can be stored as html report. Through various queries the reports can be customised to the requirements of several target groups. These reports can contain all data from a scan or only parts, for example only the results, which represent high risks for the network devices. In contrast to other network and security scanners the user is not confronted with a lot of data. NetScanAssistant enables the option to evaluate the results and to search for certain information. This makes the challenge of analysing the information from a network scan more comfortable and flexible.

3 DISCUSSION

At the moment NetScanAssistant is available as a prototype. The basic logic functions of scanning and result management work well as described but before publication it has to be tested much longer and more extensively. Tests in large networks with scan durations of several days and megabytes of data were performed successfully but the process of testing and further development has not finished yet. Also the integration of other existing applications and tools or functions for control and report management demands a new and improved security concept. The current NetScanAssistant surely can be taken as a basis but there is also a lot of work needed to be done in the future.

4 CONCLUSION

NetScanAssistant is a network scanner platform that uses state of the art network scanners to provide a user that kind of information about the devices in a network which is relevant for him. The information gained is saved in a relational database to allow a flexible report management, a high level of security, and long-term storage. At the moment an Eclipse plug-in allows to control a scan and to evaluate the results and to search for certain information with predefined or own queries (templates). The modular design makes it simple to expand the given functionality and to add by example new modules that support scan control and result evaluation via e-mail, web, PDAs, mobile phones, etc. Some query results already can be stored as html reports.

5 REFERENCES

[1] Fyodor, *Nmap security scanner*. URL, http://www.insecure.org.
[2] Deraison, R., *Nessus*. URL, http://www.nessus.org.
[3] Apache Software Foundation, *Log4j*. URL, http://logging.apache.org/log4j.
[4] JBoss Inc., *Hibernate*. URL, http://www.hibernate.org.
[5] Gamma, E. and Beck, K., *Junit*. URL, http://junit.org.
[6] Eclipse Foundation, *Eclipse*. URL, http://www.eclipse.org.
[7] MITRE Corporation, *Common vulnerability exposure (cve)*. URL, http://cve.mitre.org.
[8] Symantec Corporation, *Securityfocus*. URL, http://www.securityfocus.com.

Austrian Mobile Equipment Tracking System (AMETS)

Johann Wallinger

Secure Information Systems, Upper Austria University of Applied Sciences,
Hauptstraße 117, A-4232 Hagenberg, AUSTRIA

ABSTRACT

Internationally, mobile phone theft is increasing, but the clearing rate is low. Thus there is a pressing need of immediate action. In particular, the protection of the *International Mobile Equipment Identity (IMEI)*, the unique identifier of mobile phones, is of importance, since a manipulation of this ID entails the anonymization of the equipment which would prevent the blocking and tracking of stolen mobile phones.

Great Britain and the GSM Association (GSMA) have instituted pioneering measures regarding the instruction of, on the one hand, mobile phone manufacturers to implement more secure standards for IMEI protection and, on the other hand, to establish structures which allow for an active combat of mobile phone theft.

Austria has also realized the problems surrounding this issue. Therefore, the author has created the *Austrian Mobile Equipment Tracking System (AMETS)* which focuses on the national prevention of theft of and dealing with mobile phones. This system implies the *National Equipment Identity Register (NEIR)*, a database of stolen IMEIs which can be accessed by national mobile phone companies and the Austrian police. Mobile phone companies operate a local *Equipment Identity Register (EIR)* which is synchronized with the NEIR. Additionally, IDs of stolen mobile phones are marked in the NEIR upon login attempt so that investigators can recognize them. This has two chief effects: Firstly, the use of registered, stolen mobile phones can be prevented, and secondly the localization of illegal mobile phone users can be enabled. However, in the end the success largely depends on the integrity of IMEIs in mobile phones.

Motivation:

Austrian detectives commissioned with the investigation of mobile phone theft and related crimes currently have no adequate technical appliances to combat mobile phone related, criminal activities. Given the fact that technical instruments and data needed for tracking lost or stolen mobile phones are already available, this is a really unsatisfying situation which has motivated the AMETS project.

Goal:

The ultimate ambition of the AMETS is to provide contemporary structures and functionalities needed for abatement of theft and concealment of mobile phones. The following services are provided:

- a central database, accessible by investigators and mobile network operators,
- functions for synchronization of local databases[1] with the central database,
- state of the art protection of hosted data with emphasis on access and availability,
- functions accessible[2] by investigators to analyze IMEIs.

Contact: amets@s3c.at

---

[1] The Austrian government of the interior (BM.I) and mobile phone companies operate local databases with IMEI-lists of stolen and lost mobile phones.

[2] Via SMS-queries or requests transmitted over the Internet (SOAP-calls, Web-form)

1 INTRODUCTION

To track down stolen and lost mobile phones and prevent usage of these devices two major improvements have to be made: First and foremost stronger device security has to be established to aggravate manipulation of IMEIs and thus prevent resulting anonymization of mobile phones.

In addition structures which allow for a collaboration of mobile network operators and investigators, without violating the privacy of legitimate users, have to be established. Building up these structures and providing functionalities needed for Austrian investigators to combat crime on the mobile phone sector are tasks at hand for the project AMETS.

2 APPROACH

To allow for an active combat of theft and concealment of mobile phones many systems of different institutions have to interact with each other which implies that standardized interfaces have to be defined and trust relationships between involved systems must be established. To minimize the work needed for system interconnection a central platform is built, providing interfaces for synchronization and functionalities for analysis of essential data. This platform termed AMETS is the key component of the overall system. The following sections describe all institutions and their systems that are actively or passively contributing to the overall system. Figure 1 illustrates the entire model.

2.1 Mobile Network Operators

Only mobile network operators are able to associate the holder of the mobile phone with the actual device itself. Each device activity within the operators network is recognized and therefore certain actions depending on the status of the mobile phone can be settled. The status of mobile phones is represented through their IMEIs which are stored on either a blacklist, whitelist or greylist. A dynamic system which allows for an

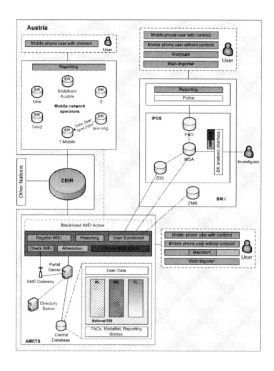

Fig. 1. Austrian model for abatement of theft and concealment of mobile phones

active tracking of stolen and lost mobile phones is realistically only possible in cooperation with mobile network operators.

Assignment

The mobile network operators task within the national mobile phone crime prevention system is on the one hand to abandon illegal mobile phones from their networks and on the other hand to mark active stolen or lost mobile phones so that investigators are able to recognize them. Furthermore, customers bound by contract should be enabled to get their stolen or lost mobile phones barred. The main duties of mobile network operators can be summarized as follows:

- daily synchronization of the local EIR with the national EIR

- IMEIs which - at the time of network logon - cannot be found on neither black- nor whitelist must be registered on the greylist.
- IMEIs which - at the time of network logon - are registered on the blacklist must be marked in a way that investigators are able to recognize them by using certain functions provided by the AMETS

2.2 Austrian Federal Ministry of the Interior (BM.I)

The Austrian BM.I is hosting, amongst others, systems and applications representing the IPOS (Integriertes Polizeiliches Sicherheitssystem [1], Figure 2) used by the Austrian police to capture (PAD - Protokollieren Akten Daten), process and query (IKDA - Integrierte Kriminalpolizeiliche Datenanwendung) and govern (ZDS - Zentrale Datensammlung; which represents a revised version of the EKIS (Elektronisches Kriminalpolizeiliches Informationssystem)) crime related data.

Assignment

The IPOS enables investigators over interfaces from the IKDA and the ZDS to the AMETS to track and analyze IMEIs of lost or stolen mobile phones. The main duty of the BM.I is thereby to provide resources needed for the operation and administration of applications and related systems.

2.3 AMETS

The AMETS (see Figure 3) is the core component of the overall system and consistently holds data provided by users of available services. For example, mobile network operators use the AMETS to synchronize their EIR, so that blacklisted IMEIs show up on all blacklists of connected networks. Furthermore the AMETS provides functionalities for investigators to effectively check and analyze IMEIs.

Assignment

In general, functionalities provided by the AMETS can be summarized as follows:

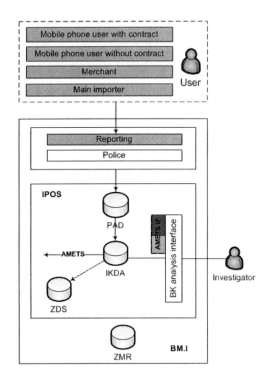

Fig. 2. IPOS

- functions that allow investigators to analyze IMEIs.
- functions for owners of mobile phones to register their devices and report them lost or stolen.
- functions for mobile network operators and the BM.I to synchronize their local EIR with the NEIR.
- synchronization of the NEIR with the Central Equipment Identity Register (CEIR) hosted by the GSMA. Due to the international character of mobile phone theft it is necessary to maintain in addition to the national blacklist an international list, which is administrated by the GSMA.
- user authorization and authentication.

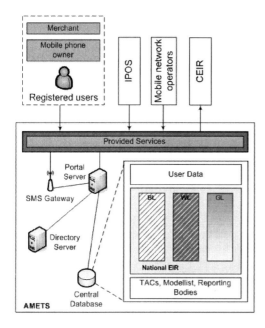

Fig. 3. AMETS

Services

Depending on the assigned user-roles the AMETS grants access to different services which are outlined below.

User Enrolment Owners of mobile phones are - after successfully completing the registration process - enabled to administrate their mobile phone related data (in particular the IMEI of the device) and use some of the functionalities provided by the AMETS. Especially, the *Check IMEI* and the reporting of lost or stolen mobile phones are useful add-ons.

Register IMEI This service is used to register IMEIs of stolen or lost mobile phones. In particular, the IPOS calls this function whenever mobile phones are reported lost or stolen. Also main importers use this function to whitelist IMEIs of imported mobile phones. Registered users are also able to register their IMEI, but since there is no guarantee that the mobile phone has (consciously or unconsciously) been lawfully obtained, all IMEIs registered by users are put on the grey list and clearance might be given after validation through local police. An exception is made when the IMEI which is to be registered is already found on the whitelist (e.g. the IMEI was registered by main importers). In this case the IMEI is associated with the user.

Check IMEI Allows for a detailed analysis of IMEIs. Besides the status of the IMEI (blacklisted, whitelisted, greylisted) other valuable information can be extracted out of this ID. The first eight digits represent the Type Allocation Code (TAC)[3] which defines the type of the mobile phone. Furthermore the last digit (check digit) is used to validate the 14+1 digit IMEI.

Depending on the user role a more or less detailed analysis report is returned to the querant. *Check IMEI* might be used by registered users to validate the status of mobile phones before actually buying them. The *Attestation* process also uses this service to check on the status of certain IMEIs.

Attestation Uses the *Check IMEI* function to discover whether the IMEI is registered and if so which list it is assigned to. If the IMEI is whitelisted or unknown to the system, a certificate is generated which can be used by merchants to attest third parties that this mobile phone was not marked lost or stolen and can be used without problems in all Austrian mobile networks (of course provided that the device is not bound to a certain operator).

List active, illegal IMEIs This service allows investigators to receive a list of IMEIs of stolen and lost mobile phones which were reported active by mobile network operators.

---

[3] In mobile phones registered before 12/31/2002 these eight digits are composed of a six digit Type Approval Code and a two digit Factory Allocation Code (FAC). Due to a numbering shortage the six digits were extended to eight digits. During a transition period which lasted two years the digits seven and eight of allocated TACs were set to zero

Reporting Enables registered users to blacklist IMEIs enrolled within their account. Once the IMEI is blacklisted its status can only be changed by the police. Since contracted users might call their service provider and ask to have their mobile phone barred, mobile network operators might use this function to blacklist certain IMEIs. It depends on the SLA whether IMEIs are blacklisted in the local EIR which gets synchronized with the NEIR once a day or are blacklisted directly in the NEIR making them available for investigators without delay.

Change IMEI Status This service allows for a status change of already registered IMEIs e.g. over the IPOS when a case is closed and the mobile phone is returned to its legal owner.

Tag active IMEIs Mobile network operators use this function whenever IMEIs upon network logon were found on the blacklist of the local EIR. When calling this function the IMEI has to be provided which then will be tagged active in the NEIR. This allows investigators to receive a list with IMEIs of currently active mobile phones reported as lost or stolen.

2.4 GSMA

The GSMA represents interests of mobile network operators world wide and plays a major role in the combat against theft and concealment of mobile phones. Amongst others it operates the CEIR, a central register with IMEIs of stolen, lost, unknown, and allocated IMEIs which can be accessed by members of the GSMA to allow for an international exchange of illegal IMEIs.

Assignment
Besides its role as representative of mobile network operators the GSMA instituted many drafts and regulatory papers regarding the security in GSM Networks. Also, the protection of IMEIs was targeted in a letter to the EU-Commissioner [2] in which they emphasize to compel manufacturers of mobile phones to establish state of the art protection mechanisms and permanently review and adopt them over time if needed.

In reference to the AMETS the GSMA provides the following functionalities:

- access to the international blacklist
- access to the international whitelist

2.5 Main Importers

The majority of mobile phones sold and used within Austrian borders are imported by national importers. Through registration of imported mobile phones a list of known good IMEIs is built which can be very useful in the IMEI validation process.

Assignment
The following tasks are to be fulfilled by main importers:

- registration of IMEIs of imported mobile phones.
- registration of identification data of supplied merchants in conjunction with IMEIs and time of delivery.

2.6 Legislative Body

Each proceeding of investigators has to be performed within legal borders which have to be adopted from time to time to meet requirements on new forms of crime emerging with rapidly changing technologies.

Assignment
To allow investigators take legal proceedings against persons who manipulate IMEIs on purpose, the Austrian STGB needs to be adopted. Currently no passage is defined which allows for lawful measures to be taken against such behavior. Great Britain already enacted a law ("Mobile Telephones (Re-programming) Act 2002" [3]) which addresses exactly this issue giving police forces a lawful basis for their proceedings against criminal activities on the mobile phone sector.

2.7 Merchants

Through the use of the *Check IMEI* function merchants are given the opportunity to validate the legal status of the mobile device. In addition, a certificate for mobile phones not found on the black- or greylist is issued by the AMETS to attest third parties that the mobile phone can be operated within Austrian mobile networks. Accredited and registered merchants have the possibility to report mobile phones as stolen.

The following services are to be used by registered merchants:

- User Enrolment
- Register IMEI
- Check IMEI
- Attestation

2.8 Mobile Phone User

Users of mobile phones must not only take an active part upon theft or lost of their device, they should pro-actively use provided functionalities like *Check IMEI* or *Register IMEI* to on the one hand profit from legality assurances and on the other hand improve the value of the overall system by providing useful information like IMEI and merchant IDs. Especially upon negative answers returned by the *Check IMEI* function it is important to know which merchant was putting up the mobile phone for sale.

The following services are to be used by mobile phone users:

- Reporting via mobile network operator
- Reporting via Police
- User Enrolment

Upon successful enrolment the following services may be used in addition:

- Check IMEI

2.9 Police

Through the connection between AMETS and IPOS police forces are enabled to use functionalities provided by the AMETS. Investigators assigned with clarification of crimes are able to use almost every functionality provided by the AMETS whilst executives responsible for recording of reported crimes only use the reporting function. Separation of this user roles is performed within the IPOS and cannot be controlled by the AMETS.

The following services are to be used by investigators:

- Register IMEI
- Check IMEI
- List active, illegal IMEIs
- Change IMEI Status

2.10 Summary

The implementation of this approach will allow for an active combat against theft and concealment of mobile phones. The core component of the overall system provides functionalities which allow for exchange and analysis of data provided by connected systems. Investigators profit from various analysis functionalities which can be accessed also by mobile investigators (SMS) allowing for in time reporting. The AMETS functions as link between mobile network operators, the police, main importers, merchants, and holders of mobile phones, and allows for an effective cooperation and interworking between all participating parties.

3 DISCUSSION

International criminal organizations, corrupt merchants and unreliable sources of information: The fight against theft and concealment of mobile phones has to be fought on many borders. The main goal is to eliminate such crime beforehand which is realistically only possible if ways can be found to prevent potential users from buying illegal mobile phones to reduce the gainable profit for thieves and oppose it to the high risk of the felony itself. Adopting the earlier described concept will help to achieve this goal since known illegal mobile phones are barred from networks of Austrian telecommunication service providers which will encourage customers to check on the legal status of their phone in advance. Besides the building of technical structures for an effective fight against crime, the creation of a general awareness regarding possible risks and countermeasures in the context of theft of mobile phones is of essential importance, since without proper reporting of offenses and utilization of provided services the effectiveness of the system is little. Whilst users of mobile phones can be convinced to use the provided services - since it is for their own good - mobile network operators might not see the value (profit) in providing crime prevention services, which leads to a major problem: Conflicts of interest. While investigators are of course keen on preventing and eliminating criminal activities on the mobile phone sector, the mobile network operators point of view is more "economical" - provided services must at least be self-liquidating.

In contrast to organizational wattles the actual technical implementation is far less problematic. Although aligned with some time and effort the realization is realistically manageable as the already implemented prototype documents. The efficiency of the total system rises and falls with the security against manipulation of the mobile phones IMEI which must be implemented at its best by mobile phone manufacturers. Rapid development of new technologies and production processes make it already possible to embed hardware security devices at moderate prices, e.g. Nokias BB5 enabled devices already use a crypto processor and its IMEI is - at the time of writing - tamperproof. At least by now a general awareness on the protection of the IMEI has been created which was not the case when the GSM was specified. In 1999 the ETSI document TS 122.016 [4] was extended with the following clause:

The IMEI is incorporated in an UE module which is contained within the UE. The IMEI shall not be changed after the ME"s final production process. It shall resist tampering, i.e. manipulation and change, by any means (e.g. physical, electrical and software).

This requirement is valid for new GSM MEs type approved after 1st June 2002. However, this requirement is applicable to all 3GPP system compatible UEs from start of production.

The IMEI is the only traceable anchor between mobile phones and subscriber.

4 CONCLUSION

The development of the AMETS gives investigators various possibilities to combat theft and concealment of mobile phones. If mobile network operators are participating, lost and stolen mobile phones can be barred out of Austrian mobile networks and investigators will be given the possibility to explore these devices. Furthermore, consumers get the possibility to check on the legal status of their mobile phone, making it harder to infiltrate the national market with stolen mobile phones. The link to the CEIR also prevents internationally stolen, imported mobile phones from being used in Austrian networks.

Current Status

The operation of the overall system is at the time not accomplishable, mainly because Austrian mobile network operators are not cooperating. Profitable discussions were aborted when it came to privacy aspects of subscribers, since operators have the obligation to observe confidentiality of subscriber information which is in their opinion likely to be violated through the indirect link between IMEI and subscriber. The fact that operators of the AMETS do not have the information to make that link - unless this information is provided by network operators which is not necessary for an operation of the AMETS - has been ignored. Based on Austrian law, the delivery of personal data of subscribers to investigators is allowed only at sight of a valid court order.

Nevertheless investigators can use analysis functionalities of the AMETS which are gaining value as it is getting harder to manipulate IMEIs of mobile phones. However, as long as mobile network operators are not participating, every successful proceeding of investigators is random and will not considerably change the current situation.

5 REFERENCES

[1] Redaktion "kripo-online.at" (2004), *Project ipos, interview with mr. karl pogutter.* Internet Resource, http://www.kripo-online.at/krb/show_art.asp?id=762.

[2] GSM Europe (2002), *Letter to the eu-commission to adopt the rtte directive.* Internet Resource, http://www.gsmworld.com/gsmeurope/documents/pl23/letter_liikanen_mob_theft_final_101202.pdf.

[3] UK Parliament (2002), *Mobile telephones (re-programming) act 2002.* Internet Resource, http://www.opsi.gov.uk/acts/acts2002/20020031.htm.

[4] ETSI 3rd Generation Partnership Project (3GPP) (2005), *Digital cellular telecommunications system (phase 2+); universal mobile telecommunications system (umts); numbering, addressing and identification (3gpp ts 23.003 version 6.8.0 release 6).* Internet Resource, http://www.3gpp.org/ftp/Specs/html-info/23003.htm.

Secure Token Concept within a Near Field Communication Ecosystem[*]

Gerald Madlmayr[a], Oliver Dillinger[a], Sebastian Gierlinger[b], Peter Kleebauer[b], Roland Pucher[b], Dieter Vymazahl[b], Josef Langer[a], Christoph Schaffer[a], Jürgen Ecker[b]

[a] Upper Austria University of Applied Sciences—Research and Development Competence Center, Hauptstraße 117, A-4232 Hagenberg, AUSTRIA
[b] Upper Austria University of Applied Sciences—Secure Information Systems, Hauptstraße 117, A-4232 Hagenberg, AUSTRIA

ABSTRACT

Near Field Communication (NFC) enables fast and easy wireless communication between devices within the range of about 10 centimeters. The technology can be used for exchanging small amounts of data for payment systems or the configuration handling for other technologies like WiFi or Bluetooth. NFC can also make use of a secure element to store sensitive data, it is an optimal technology platform for contact less electronic access and payment solutions. The secure element offers a dual interface, so that it can be accessed on one side through the base band of the handset and on the other side through the contact less interface of the NFC module. Besides storage capacity of an arbitrary amount of memory the secure element also comes with cryptographic functionality, which enables us to encrypt transactions through the contact less interface or over the base band.

We show, how a mobile device can be used as a secure token to access buildings, parking lots or even serve as an event ticket. Thus, we present a client-server system in combination with NFC devices in order to realize a secure communication and token exchange. The main focus of this paper is the realization of a secure protocol that enables different parties in a public key infrastructure to securely exchange keys and tokens.

As NFC is compatible to already existing standards like *Philips' Mifare*[1] or *Sony's Felica*[2], the reader infrastructure does not need to be changed and so NFC can easily be integrated into already existing solutions. NFC devices that are capable of card emulation will definitely make mobile computing and commerce applications more convenient for the user and will help to save costs and time.

1 INTRODUCTION

NFC is an amendment to existing contactless smart card systems but still is compatible with them. Those are ISO 14443-A-compliant systems like Philips' Mifare as well as Sony's proprietary Felica. NFC devices can operate in three different modes:

- *Reader/Writer:* The NFC device acts as a smart card reader/writer and can access information stored on an RF-transponder.

---

[*]This work is sponsored by FFG (Austrian Research Promotion Agency), Project #811408.

[1] http://www.semiconductors.philips.com/products/identification/mifare/
[2] http://www.sony.net/Products/felica/csy/jre.html

- *Peer-To-Peer:* The NFC device establishes a link to another NFC device for bidirectional data exchange.
- *Tag emulation:* The NFC device acts as a smart card. A proximity card reader can access information stored in the NFC device. This mode is important in order to integrate NFC devices into already existing legacy infrastructures.

The *tag emulation mode* is a promising alternative to ordinary smart cards or RF-transponders. The information presented by the NFC device to the reader is stored in a secure element that is also part of the NFC device. The secure element is an integrated circuit (IC) that is built to store sensitive data. Access to the information is only granted if the correct keys for reading or writing are known. The secure element of an NFC device can be accessed over the base band and the RF-interface.

The system presented in this paper makes use of this dual interface chip to realize a secure token concept. Ticketing applications, a special form of tokens systems, are one of the services in m-commerce mobile phones could be used for. A ticket is a proof indicating that its holder has paid for or is entitled to use a specified service or right [1]. Today there are various forms of ticketing systems using simple paper slips, barcode systems or RF-tickets.

2 RELATED WORK

Today most of the tickets are still printed paper slips. Paper tickets are cheap and simple to produce. Normally, tickets are directly produced by the ticket issuer. Another option is that the consumer receives a digital document online that has to be printed by himself/herself.

A newer possibility is to obtain a ticket via short message service (SMS)[3]. In this case the ticket is bought via web or WAP. Payment can be performed by credit card or by the billing system of the network operator. After a successful transaction, the user receives a unique ID in an SMS that serves as a link to the server sided stored ticket [2].

Another version of SMS ticketing to send visual instead of textual information using MMS. In this case the ticket vendor sends a unique ID encoded as a barcode or data matrix [3] to the handset, which is presented at the entrance gate [4], [5]. When using SMS ticketing services, the information concerning the rights ensured by the ticket are stored on a central server. For verification of the ticket an online connection to the central server is necessary in order to check if the ticket is valid.

Within the public transport sector smart cards store the tokens. The user obtains a smart card and then recharges it with tickets or values for the public transport system. The access is granted if the smart card contains a valid ticket. The access system in this case is distributed and the verification of the ticket is done directly at the point of entrance without a server connection. In order to do so, the smart card must be able to store the ticket in a secure place within the card [6].

Also mobile devices with integrated technology for local communication have been considered to be part of a ticketing application. [7] and [8] propose systems that also deal with security issues of distributed digital ticketing systems. These systems already give a sound idea of how such systems could be realized, but do not consider usability and administrative facts of distributed ecosystems.

3 TOKEN APPLICATION
3.1 Architecture

Our concept for a token application based on NFC consists of four different parties (Fig. 1) communicating in a secure way by relying on a X.509 Public Key Infrastructure (PKI) [9].

- *Token Distributor:* The token distributor is the trusted authority in this system. This instance signs the key of the other parties and

---
[3] Also known as *short mail* in Japan.

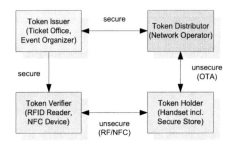

Fig. 1. General Architecture of the System.

Fig. 2. Public Key Infrastructure.

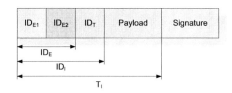

Fig. 3. Structure of a token.

is responsible for the personalization of the token issuer and the token holder. The token is received from the issuer through IP[4] and is sent to the holder over the air via SMS or Wap-Push.

- *Token Issuer:* The token issuer is responsible for cutting the tokens and sending them to the distributor. The distributor and the issuer are connected through IP.
- *Token Holder:* The token holder receives the token over the air and stores it in a secure store until it is handed over to the verifier over the air interface (NFC).
- *Token Verifier:* The token verifier reads the token from the token holder and verifies the authenticity of the token.

In such a distributed system the threats to the application are much higher than in a centralized system. Therefore the following security related issues have to be considered:

- Transfer of the token to the consumer
- Authenticity of the token issuer
- Validation of the tokens
- Authenticity of the Reader
- Relay- and Replay-Attacks

---

[4] Internet Protocol (IP) is a data-oriented protocol used for communication across packet-switched networks.

The OTA and RF transactions are the most critical ones in this system, as these channels do not support secure communication yet. The OTA transaction must be cryptographically secured and the authenticity of the parties evolved must be ensured. Also the communication over the RF interface is a critical one. Thus both the token verifier and the token holder must conduct a bilateral authentication.

3.2 Public Key Infrastructure

To secure the architecture shown above we propose a PKI (Fig. 2). The top level instance in this system is the token distributor that acts as a trusted authority (TA) and can issue self-signed certificates. Thus the token distributor can sign the public keys of the token issuer and the token holder. This is a prerequisite for the instances to participate in the ecosystem and to demonstrate their authenticity.

3.3 Components

3.3.1 Token Definition: The token (Fig. 3) containing access rights or other payload information is divided into the following data fields:

Field	Description
ID_{E1}	ID of event set by the issuer
ID_{E2}	ID of event set by the distributor
ID_E	ID of the event
ID_T	Unique ID of the token itself per event
ID_I	Unique ID of the token
T_I	Token of the issuer
Payload	Information within the token
Signature	Signature of the distributor (56 bytes at ECC with 193 bits)

Table 1. Explanation of fields used in the token.

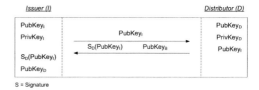

Fig. 5. Key Exchange between the issuer and the distributor.

Fig. 6. Key Exchange between the holder and the distributor.

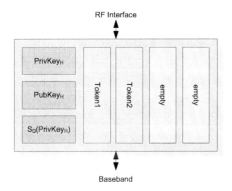

Fig. 4. Secure store of the token holder.

The size of the token should not exceed 140 bytes, as this is the amount of data that could be sent with one SMS. Sequential Chaining for SMS is possible, but would make the system more complicated and require additional security components.

3.3.2 Secure Store (Token Holder) The secure store (Fig. 4) of the token holder is a place where the public and private key of the holder and also the signature of the public key created by the distributor is stored. The second function of the secure store is to hold the tokens between the delivery through the base band and the communication over the air interface. The secure store in an NFC device can be realized by using the secure element.

3.4 Protocol

The protocol has to handle the following parts of the token usage process.

- *Key Exchange (Initialization)*
- *Preparation of Token (Issuing)*
- *Booking of Token (Distribution)*
- *Verification of Token*

3.4.1 Key Exchange (Initialization) In the initialization phase the token handling application is installed and the secure store of the token holder is initialized. Each Instance of the PKI generates its public and private keys. The token issuer and the token holder send their public keys to the token distributor that signs these keys with its private key and send the signature back. To verify the signature the issuer and holder also obtain the public key of the distributor (Fig. 5 and 6).

3.4.2 Preparation of Token (Issuing) The preparation of the token is also split into two parts. In the first step the issuer sends its part of the event ID to the distributor. The distributor completes the event ID, signs the whole event ID with its

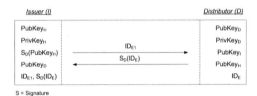

Fig. 7. Generation of event ID with signature of distributor.

Fig. 9. OTA transfer of the token to the holder.

Fig. 8. Hand over of the event ID to the verifier.

private Key and sends it back to the issuer. The issuer can now verify the signed event ID with the public key of the distributor obtained during the key exchange (Fig. 7).

In the second step the issuer hands over the signed event ID and the public key of the distributor to the verifier. The verifier is now able to check the tokens needed for an event with the ID obtained (Fig. 8).

3.4.3 Booking of Token (Distribution) First the issuer completes the fields of the token mentioned in table 1. Then he signs the token with its private key and sends it, together with the signature of the public key, to the distributor. By using the public key of the issuer the distributor can verify the authenticity of the signature of the token. Before distributing the token to the holder the distributor checks if the event ID is valid and not black-listed.

As the communication over the air is not secure the distributor encrypts the token with the public key of the holder. Therefore, only the authorized holder is able to decrypt the token. The encrypted package is signed with the private key of the distributor. With the public key of the distributor

the holder is able to check the signature of the encrypted token and check if it comes from a secure source. If this is the case, the token is decrypted with the private key of the token holder and stored plain in the secure store of the holder. Storing the token plain is not a security issue, as the secure store itself is save (Fig. 9).

3.4.4 Verification of Token At the point of verification the verifier is polling for token holders. Therefore the verifier first sends the signed event ID. The holder can verify the authenticity of the verifier by checking the signature of the event ID. If the holder does not hold an appropriate token the communication is stopped. Otherwise, the verifier sends a random number, which is going to be part of the session key later on.

Secondly, the holder hands over its public key and the signature of the distributor for its public key. This is the proof of authenticity of the holder to the verifier. The holder also adds a second random number in order to contribute to the session key used later.

In the third step the verifier encrypts the message and both random numbers with the public key of the holder and sends both random numbers back. Only the holder with the correct private key can extract the correct random numbers and set up a session key for a secure communication.

Both parties have now agreed upon a session key that allows the verifier and the holder to exchange the token in a secure way. The verifier can validate the payload of the token and set

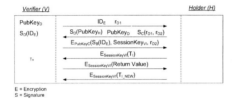

Fig. 10. Protocol flow at the point of verification.

an appropriate action for the token holder. If necessary the token is erased from the secure store of the token holder (Fig. 10).

4 CONCLUSION

The proposed system enables a secure token handling in a distributed NFC environment. By introducing a trusted authority, different service providers can offer their services within a secure system.

Besides the trusted authority also a secure store in the form a secure element at the ticket holder is necessary. Unfortunately, the NFC-Forum[5] marked out that such a secure element is not mandatory for an NFC device. But so far, all available NFC devices suitable for ticketing and payment are equipped with such a secure store.

Finally the industry has to agree on a common protocol for the communication between those parties and on a unique ticket format. This seems to be the most challenging part, as there are already numberless formats of tickets defined.

5 REFERENCES

[1] Webster's Encyclopedic Unabridged Dictionary of the English Language. Random House, New York, NY, USA.

[2] Ching, L. T. and Garg, H. K. (2002) Designing SMS applications for public transport service system in Singapore. *ICCS 2002. The 8th International Conference on Communication Systems, 2002*, 11, vol. 2, pp. 706–711, Nat. Univ. of Singapore, IEEE Computer Society.

[3] International Organization for Standardization (2004) *ISO/IEC16022 - International symbology specification – Data Matrix*. International Organization for Standardization, 1 edn.

[4] France Telekom (2004), m-Ticketing Services. http://www.francetelecom.com/sirius/rd/en/ddm/en/technologies/ddm200405/techfiche4.php, online; accessed 05/12/2005.

[5] ELCA, S. (2004), SecuTix. www.secutix.com/resources/product_mobile_de.pdf, online; accessed 05/12/2005.

[6] Klaus Finkenzeller (2003) *RFID Handbook*. Carl Hanser Verlag.

[7] Byung-Rae, L., Kim, T.-Y., and Sang-Seung, K. (2001) Ticket based authentication and payment protocol for mobile telecommunications systems. *2001 Pacific Rim International Symposium on Dependable Computing*, vol. 6, p. 218, IEEE Computer Society.

[8] Feng, B., Anantharaman, L., and Deng, R. (2001) Design of portable mobile devices based e-payment system and e-ticketing system with digital signature. *ICII 2001, International Conferences on Info-tech and Info-net Proceedings*, vol. 6, pp. 7–12, IEEE Computer Society.

[9] Wolfgang Ertel (2001) *Angewandte Kryptographie*. Carl Hanser Verlag.

---

[5] http://www.nfc-forum.org

Analysis of Austrian Citizen Cards

Martina Heiligenbrunner [a,*], Daniel Slamanig [a], Christian Stingl [a]

[a] Department of Medical Information Technology, Carinthia University of Applied Sciences, Primoschgasse 10, A-9020 Klagenfurt, AUSTRIA

ABSTRACT

Motivation: The efficiency of complex communication processes can be considerably improved by the electronic transmission of data. The communication in the sensitive area of health care or E-Government requires mechanisms to guarantee secure data transmissions. Today, this can be realized by the use of cryptographic methods like digital signature and data encryption. Generally, smart cards are used for the storage of key information because they are the best practice to provide security. The concept of the Austrian citizen card meets all these requirements and can be considered as a signature and encryption card for the use in communication in sensitive areas. This work focuses on the following two main types of citizen cards: a.sign Premium delivered by the company A-Trust and the e-card delivered by the Main Association of Austrian Social Security Institutions (MASSI).

Results: These different types of citizen cards in Austria use two kinds of public key methods - RSA-Cryptosystem and Elliptic Curve Cryptography (ECC). Furthermore, the cards contain at least two different types of certificates - the qualified and the simple certificate. The use of the qualified certificate for creating secure signature has to be performed by software that is compatible to the Security Layer of the citizen card definition. Such software is already available for both RSA and ECC algorithms. Using the simple certificate in Microsoft operating systems to sign or decrypt documents, a Cryptographic Service Provider (CSP) for the applied algorithms (RSA, ECC) needs to be integrated in the system. A CSP for the RSA algorithm is already available, but the ECC algorithm is not supported yet. However it will soon be available in the form of a plug-in for Microsoft Windows and will be integrated in the next generation of Microsoft Operating Systems. The Public Key Infrastructures (PKIs) which are necessary for the management of the certificates show weaknesses concerning the unique identification of certificate holders. As soon as CSPs are available for ECC algorithms and the unique identification of the certificate holders can be guaranteed, these types of citizen cards can be considered as instruments that improve the security in Information Systems significantly.

Contact: Martina.Heiligenbrunner@fh-kaernten.at

1 INTRODUCTION

The efficiency of complex communication processes can be considerably improved by the electronic transmission of data. The communication in the sensitive area of health care or E-Government requires mechanisms to guarantee secure data transmissions. A secure data transmission is mainly characterized by four security properties:

1. Authenticity
2. Integrity
3. Confidentiality
4. Non-repudiation

---

*to whom correspondence should be addressed

These security properties for electronic transmission can be realized by the use of cryptographic methods like digital signature and data encryption. For the storage of the key information, smart cards are generally used because they are the best practice to provide security. The Austrian citizen card describes a concept which enables every Austrian citizen to participate in processes in E-Government. This work focuses on following two main implementations of citizen cards: a.sign Premium delivered by the company A-Trust and the e-card delivered by the MASSI. The first provides

- a.sign Premium: A qualified certificate for secure digital signatures.
- a.sign Premium encryption: A simple certificate for data encryption and simple digital signatures.

and the second one

- e-card administrative signature: A simple certificate for administrative digital signatures.
- e-card simple signature: A simple certificate for data encryption and simple digital signatures.

The card a.sign Premium has to be purchased and a yearly fee has to be paid. The e-card which is the substitution of the health insurance vouchers requires no additional fee and the certificates can be activated via the Internet free of charge. For other available implementations of the citizen card see for instance [1]. Table 1 illustrates the approximate amounts of available cards and activated certificates. Using digital signature and data encryption for the protection of personal data, legal circumstances have to be carefully considered. In Austria the protection of personal data in electronic communication processes is regulated by several laws.

1.1 Legal Framework

1.1.1 Federal Act concerning the Protection of Personal Data 2000 The Federal Act concerning the Protection of Personal Data [2] is the law that is currently in force in Austria for data protection and, therefore, it is the most important legal provision for the protection of data in Austria. Basically, it defines the fundamental right of every citizen to data protection and regulations that exist when using personal data regarding interests in secrecy deserving protection.

1.1.2 Federal Electronic Signature Law The Federal Electronic Signature Law [3] with its regulation [4] is based on the European law (Directive of the European Parliament and of the Council on a Community framework for electronic signatures). Besides administrative and security-related objectives, both parameters for technical methods and components that are considered as secure according to the state-of-the-art of technology, as well as methods for technical components are specified.

1.1.3 Regulation for Administrative Signatures The regulation for Administrative Signatures [5] clarifies security-relevant and organization-relevant questions that arise in relation to the administrative signature. In contrast to the secure signature, this is a temporary and simplified variant for the use in the field of E-Government and is equivalent to the secure signature until the end of the year 2007.

1.1.4 E-Government Act The E-Government Act [6] regulates the legal framework that needs to be considered when realizing an extensive E-Government concept. The main aspects of this law concern the identification and authentication in electronic business with public bodies as well as the use of the citizen card in the private sector.

1.1.5 Health Telematics Act The Health Telematics Act represents article 10 of the Health Reform Act 2005 [7]. It clarifies essential questions in relation to the use of Information and Communication Technology (ICT) in the field of health care. The law defines the framework for the use of digital signatures to keep the integrity of health related data and the use of encryption methods to protect the confidentiality of these data.

Provider	Product	Start	Cards	Certificates
A-Trust	a.sign Premium1[a]	Jan. 2003	17.700	17.700
A-Trust	a.sign Premium2[b]	Jan. 2005	39.600	39.600
MASSI	e-card	Sep. 2005	8.200.000	5.500

[a] Austrian citizen card provided by the company A-Trust

[b] Austrian citizen card provided by A-Trust or other companies (e.g. Maestro card, credit cards)

Table 1. Amount of cards and certificates until July 2006

Based on this legal situation, the so-called citizen card concept was developed in Austria.

2 CITIZEN CARD CONCEPT

The Austrian citizen card concept ensures a secure handling of electronic data traffic in governmental and administrative business processes through the application of cryptographic methods. The core part of this concept is the so-called citizen card describing a functional concept that is independent of the technical implementation. This functional concept includes two principal elements:

1. Authentication: Every person can participate in electronic business processes and provide right-conformal signatures using digital signatures.
2. Identification: The cryptographic key information (public key) that is necessary for the verification of digital signatures is clearly assigned to a (natural) person by the identity link.

This allows every Austrian citizen to electronically carry out governmental business processes using an implementation of the citizen card (e.g. a.sign Premium). Concrete implementations of citizen cards can be realized in the form of smart cards, mobile phones, USB token, etc.

2.1 Digital Signature

Based on the Federal Electronic Signature Law and on the E-Government Act, the following types of digital signatures can be identified. These primarily differ in terms of security requirements and in the purpose of use:

1. Secure Signature: A secure digital signature is 100% on an equal footing with the handwritten signature. The private key thereby applied may solely be used for the signature creation.
2. Administrative Signature: In the field of E-Government, the administrative signature is on an equal footing with the secure signature until 12-31-2007. Such a signature can be created with the administrative certificate on the e-card.
3. Simple Signature: The key information, which is assigned to the simple certificate, is suitable for electronic business cases which require low to medium security. The private key can be used for signature creation, authentication, and key agreement, whereas the use of the public key for data encryption is also intended.
4. Official Signature: The official signature is realized by using a specifically characterized signature certificate and links a document to the authority it originates from.

In order to be able to verify a digital signature, the following three aspects must be considered:

1. Identification: The signatory must be uniquely identified. Within a Public Key Infrastructure which is based on the basis of X.509 (PKIX) this is mainly realized by means of the Distinguished Name (DN) [8, 9].

2. Signature verification: Examination whether the signature can be assigned to the document and the signatory.
3. Certification path: Verification of the signature certificate using a specific validation model [10, 11].

For this reason, especially in the case of using PKIX, the content of the DN, e.g. last name, first name, organization, address, etc., is of enormous importance for the unique identification of the signatory.

2.2 Identity Link

The identity link is a concept that establishes a relation between the cryptographic key information of one or more certificates and a (natural) person. This is realized through an XML data structure which includes personal data (last name, first name, date of birth), the public key information of the certificates, the source identification number (sourcePIN) of a person and the signature of the sourcePIN Register Authority. The last-mentioned signature guarantees the integrity of data and identifies the issuing authority. It has to be regarded as problematic that the certificate of the issuing authority is only valid until 02-19-2007. The identity link serves the automated identification within the field of E-Government and is deposited authentically and encrypted on the citizen card.

In Austria each person can be identified by a unique source identification number. For all Austrian citizens this number corresponds to the registration number in the Central Register of Residents (CRR number) and for all other natural persons to the registration number in the Supplementary Register (SR number). The sourcePIN is a unique ID assigned by the sourcePIN Register Authority and can be calculated from the CRR number or the SR number (base number) by applying the Triple-DES as follows [12]:

1. Initial data: Base number (12 decimal places).
2. Conversion to binary notation (5 byte).
3. Enlargement of the calculation base to 128 bit (16 byte) by means of the following format *base number base number seed base number*. Seed is a secret 8-bit value which is solely known to the sourcePIN Register Authority.
4. The binary representation is encrypted by using Triple-DES in CBC mode. The encryption key used for this purpose is secret and only known to the sourcePIN Register Authority.
5. Base64 coding of the result (includes coding in ASCII).

This sourcePIN is, as already mentioned before, adopted for the unique and automated identification in the field of E-Government. However, if it would be used as an identifying feature (e.g. as primary key) in different applications, those data could be linked. Therefore, sector-specific personal identifiers (ssPINs) are used in different sectors of State activity (e.g. building and living, health, etc.). The ssPIN can be derived from the sourcePIN using the following algorithm [12]:

1. Initial data:
   - Source identification number, Base64 encoded
   - Sector: ISO-8859-1 character sequence of the abbreviation of the sector according to the regulation of the Office of the Federal Chancellor (usually a few characters in capital letters)
2. Concatenation of sourcePIN, "+" (as character), URN-prefix, and sector abbreviation.
3. Over the resulting character sequence (the resulting string) a hash value is calculated using the SHA-1 algorithm.
4. The resulting 160 bit number can be directly used for program-internal purposes. In the case of transmission or in written form, this number has to be encoded in Base64.

Since it is possible to derive all sector-specific personal identifiers from the sourcePIN, the secrecy of it is absolutely mandatory. The certificate for signature creation and the identity link are deposited on an implementation of the citizen card. For this purpose, almost only smart cards are used at present.

In addition to the signature certificate, generally a further certificate for data encryption is available. If the same certificate is used for the signature creation as well as the data encryption, attacks on the signature procedure are possible [13].

3 SMART CARDS

Generally, smart cards offer advantages within the field of authentication because they offer a two-factor-authentication (possession of the card and PIN) and are equipped with a programmable as well as a volatile memory and a microprocessor [14].

In Austria about 8.2 million smart cards in form of a health card (e-card) were handed out to all Austrian citizens with a social insurance in the year 2005. The e-card can be used as a citizen card; however, this functionality must be activated explicitly, as mentioned above, over the Internet. This activation is free of charge. At present, the processor chips of two manufacturers, Philips and Infineon, are used on the e-card; those can be considered as almost equivalent regarding their technical specification. Both chips include a cryptographic coprocessor which supports the RSA method up to 2048 bit and ECC. Furthermore, the EEPROM of both chips has a size of 36 kByte. This memory contains the signature certificate, the encryption certificate (simple certificate), a further certificate for the application "replacement of health insurance vouchers", for each certificate the associated private key information within a protected area, and the (three) certificates of the issuing Certification Authorities (CAs). Considering also the XML data structure of the identity link and the public area (personal data and insurance data), the memory capacity for additional data is extremely limited. Therefore, these smart cards are used primarily as key cards. Since these smart cards have very limited performance and memory resources, the possible cryptographic algorithms must be specifically examined.

3.1 Cryptographic algorithms

For the creation of digital signatures and data encryption, both RSA and ECC algorithms are applied (see Figure 2). Comparing these two algorithms, the following advantages of the RSA method over the ECC method can be assessed:

- Many years of practical use
- High availability in operating systems and applications
- High performance when verifying signatures and encrypting data (with the appropriate choice of the public exponent)

Disadvantages of the RSA method over ECC are:

- Much larger key lengths ($\geq$ 1536 bit versus $\geq$ 192bit)
- Lower performance when creating signatures and decrypting data
- More storage space for certificates is required

In consideration of the advantages concerning performance and key length, ECC has been used more often than the RSA method on smart cards for the last few years. For the signature of the end entity certificates and the intermediate certificates both A-Trust as well as MASSI presently use SHA-1 as their hash algorithm. Regarding the significant improvements of attacks against the SHA-1 algorithm [15] (collision attack) the whole system could thereby be lastingly compromised in the future (through preimage attacks).

4 PUBLIC KEY INFRASTRUCTURE

The infrastructure for issuing, administrating, revoking, etc. of certificates, is called a Public Key Infrastructure (PKI). Generally, the architecture of a PKI is implemented hierarchically or meshed. In case of a hierarchical PKI this corresponds to a tree structure and consists of a root certificate, several levels of intermediate certificates and end entity certificates, whereby

Provider	Product	End Entity Certificate	Intermediate Cert.	Root Cert.
A-Trust	a-sign Premium1	RSA-1024	RSA-2048	RSA-2048
A-Trust	a-sign Premium encryption1	RSA-1024	RSA-2048	RSA-2048
A-Trust	a-sign Premium2	ECC-192	RSA-2048	RSA-2048
A-Trust	a-sign Premium encryption2	RSA-1536	RSA-2048	RSA-2048
MASSI	e-card administrative signature	ECC-192	RSA-3072	RSA-4096
MASSI	e-card simple signature	ECC-192	RSA-3072	RSA-4096

Table 2. Summary of used algorithms and key lenghts of analyzed products

each certificate (exception: root certificate) is signed by the certificate superordinated in the hierarchy. Both PKIs (A-Trust and MASSI) are built hierarchically and consist of three levels (root certificate, several intermediate certificates, and end entity certificates). Because of the small number of levels, the examination of the certification path can be accomplished efficiently. Additionally, LDAP directories and OCSP responders are available.

For the identification of the end-users in the PKI usually the Distinguished Name (DN) and specific extensions (e-mail, etc.) are used. The following user-specific information is available:

- A-Trust: Title, first name, last name, serial number, date of birth (optional) and e-mail (optional). In certain cases, a job title (e.g. lawyer) and a unique code (e.g. lawyer code) can be integrated [16].
- MASSI: First name, last name, e-mail (optional) [17]

Because of the few personal contents of the certificates, the up-to-dateness can be generally ensured over the entire period of validity of the certificate. At present, a new certificate and a new card are required in the case of a change of the personal information. This is, on the one hand, connected with relatively high costs and on the other hand it complicates the administration of the certificates for the end-users, since these must be historicized. It would be advantageous for the system if certificates could be updated, i.e. create a new Certificate Signing Request (CSR) [18] with the public key information (that was not compromised) stored on the card and the updated personal information and sign it by the intermediate certification authority.

Theoretically, each applicant can be uniquely identified using the serial number of the certificate and/or the serial number in the DN. In practice, this must be done based on the user-specific information. If neither the signatory's date of birth nor the e-mail address is known, a unique identification in the case of equality of names is impossible. In this context, two cases of name equality are differentiated:

- Variant 1: If there are two certificates where the first and the last name of the applicants are the same, we call these two persons name-equal. An applicant who possesses n valid certificates is counted n times (e.g. Foo Bar with foo.bar@provider1.at (1), Foo Bar with foo.bar@provider1.at (2) and Foo Bar with f.bar@provider2.at (3) are all considered as name-equal)
- Variant 2: The first name, last name, and the e-mail address (if available) are the same. An applicant who possesses two certificates with equal e-mail addresses is regarded as identical and will not be counted and with different e-mail adresses will be counted twice: (1) and (2) are regarded as identical and (1) and (3) as well as (2) and (3) as name-equal.

Based on this convention, the available LDAP directories of the company A-Trust and the MASSI were evaluated. Thereby, the ratio between the cumulative sums of variant 1 and

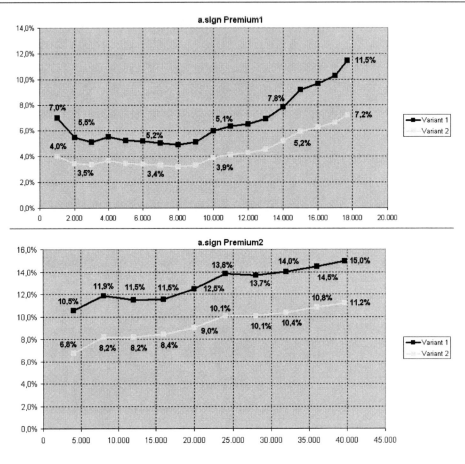

Fig. 1. The ratio of applicants which are name-equal in the A-Trust PKI (x-coordinate: number of certificates).

variant 2 and the cumulative total sum was calculated. During this calculation, the upper limit was increased by 1000 certificates with each successive step. In the case of the MASSI, revoked certificates were not considered. Due to the relatively small number of certificates (see table 1) issued by the MASSI, the ratio of variant 1 and variant 2 amounts to approximately 1.7% (see Figure 2. With the product a.sign Premium1 the result is 11.5% versus 7.2% (see Figure 1) and with a.sign Premium2 about 15% versus 12% (see Figure 1). In the latter case, this corresponds to approximately 5900 certificates of variant 1 and 4400 certificates of variant 2. Therefore, 4400 persons are in the system who cannot be uniquely identified without any additional information. With an increasing number of users, the automated identification of these gets complicated or even impossible in many cases. For a minimization of this problem, additional identification-supportive information (not compromising) of the applicant could be supplemented (e.g. organization and organizational unit).

In addition, it must be considered that the intermediate certificates of the company A-Trust are valid until December 2008. For these certificates the following scenarios are possible:

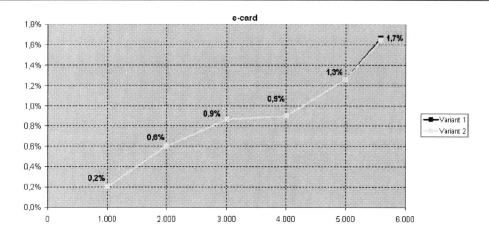

Fig. 2. The ratio of applicants which are name-equal in the MASSI PKI (x-coordinate: number of certificates).

- Extension of the validity period and rollout of the new intermediate certificates
- Reissuing of a certificate based on the same key information
- Issuing of certificates based on new key information

Depending on the selected scenario and the used verification model[11], problems with the verification of valid signatures can occur. Since, in general, the verification model is determined by client-sided configuration in the application, for example, the following unwanted effects can occur:

- If the modified shell model or the chain model is used, the complete history of the intermediate certificates must be available for the verifying person.
- If the shell model is used for the verification of a signature, no positive validation is possible after the expiration of the validity of the intermediate certificate.

For the verification of signatures additionally, the validity period of the user certificates must be considered.

5 INTEGRATION

In order to establish the security properties for applications with sensitive data, generally digital signatures and encryption are used. Digital signatures must be differentiated in secure signature/administrative signature and simple signature. The secure signature can presently only be created with a software product which implements the so-called Security Layer [19]. This environment for the creation of digital signatures is also called a citizen card environment (CCE). When creating a signature, the private key information corresponding to the signature certificate of the citizen card is used. For the activation of this information the input of a PIN (6-8 and 4 numeric places respectively) is necessary. For the verification of a signature a CCE is needed, too.

The private key information corresponding to the simple certificate can be used on the one hand for the creation of a simple signature and on the other hand for the decryption of encrypted data. In general, for both cryptographic operations a Cryptographic Service Provider (CSP) is needed in Microsoft Operating Systems. A CSP must implement the so-called CryptoAPI. This interface contains functions for the administration of certificates, encryption methods, signature methods, etc. If an appropriate CSP is available, the cryptographic operations can be easily integrated in software products (e.g. Microsoft Office, Information Systems, etc.). In the current Microsoft Operating Systems CSPs are integrated that include, among other things, the RSA algorithm; however, ECC is not supported. In the CryptoAPI next generation (CNG) which extends the CryptoAPI architecture and will be integrated in Microsoft Vista/Longhorn, ECC will be available by default. Additionally, a CSP which supports ECC, will be available in the near future (end of the year 2006) in the form of a plug-in for current Microsoft Operating Systems.

For the verification and the encryption, a standard CSP of a current Microsoft Operating System can be used. A necessary prerequisite is the support of asymmetric and symmetric methods by both CSPs.

For the A-Trust citizen card, a CSP (a-sign client) was implemented which can be used for signature creation and decryption. For the e-card, for which ECC was proposed, no CSP is available at present. This current disadvantage will be eliminated with the plug-in and with Microsoft Vista/Longhorn. Despite of this, the long-term use of the administrative signature as a secure signature cannot be guaranteed because it is only on an equal footing with the secure signature until 12-31-2007 [6].

For the digital signature of e-mails the simple certificate can be used. It must be noted that in this case the e-mail address of the applicant must be integrated in the certificate. Optionally, also more than one e-mail addresses can be integrated in the certificate. However, the use in connection with usual e-mail clients is afflicted with problems, because they verify the e-mail adress in the certificate against the used smtp account. In the case of a change of the e-mail address, a new certificate must be issued. Alternatively to that, in Microsoft Outlook the verification of the e-mail address can be deactivated [20].

6 CONCLUSION

Since the beginning of 2002 implementations of citizen cards are available in Austria. Also implementations of CCEs are provided free of charge but the handling is problematic for a technically not experienced citizen. This also reflects in the relatively small number of operations carried out in the field of E-Government. Examining the integration into existing applications, this can be done easily by means of simple certificates of A-Trust. Thus, the creation of digital signatures and data encryption can be performed. At present, this fact is not realizable in the case of the e-card. However, this deficiency will be corrected in the near future. Furthermore, it remains unclear whether, starting from 2008, administrative signatures can still be used as secure signatures or a change in the law situation will be executed. An advantage of the use of the e-card is that no additional expenses are incurred. In reference to the conception of the PKIs it must be stated that, because of the little personal information in the certificates, the unique identification of all persons cannot be ensured. The number of those persons, who are not uniquely identifiable, is about 1.7% (92 persons) in the MASSI PKI and in the A-Trust PKI about 7.2% (1280 persons) in case of a.sign Premium1 and 11.2% (4400 persons) concerning a.sign Premium2. With the further increase of applicants, this problem aggravates and the automated identification of these gets complicated. Therefore, it is necessary to include additional personal information in the certificates. If the above problems are resolved in the future, then the implementations of citizen cards analyzed in this work can be integrated more easily into existing applications.

7 REFERENCES

[1] *Die österreichische Bürgerkarte.* http://www.buergerkarte.at/.

[2] *Bundesgesetz über den Schutz personenbezogener Daten - Datenschutzgesetz 2000.* http://www.dsk.gv.at/

dsg2000d.htm.

[3] *Bundesgesetz über elektronische Signaturen - Signaturgesetz SigG 1999.* http://www.signatur.rtr.at/de/legal/sigg.html.

[4] *Verordnung des Bundeskanzlers über elektronische Signaturen - Signaturverordnung SigV.* http://www.signatur.rtr.at/de/legal/sigv.html.

[5] *Verordnung des Bundeskanzlers, mit der die sicherheitstechnischen und organisationsrelevanten Voraus-setzungen für Verwaltungssignaturen geregelt werden (VerwSigV).* http://www.signatur.rtr.at/de/repository/legal-verwsigv-20040415.html.

[6] *Bundesgesetz über Regelungen zur Erleichterung des elektronischen Verkehrs mit öffentlichen Stellen - E-Government-Gesetz - EGovG.* http://www.cio.gv.at/egovernment/law/.

[7] *Gesundheitsreformgesetz 2005.* http://www.bmgf.gv.at/.

[8] Charlisle Adams, S. L. (2003) *Understanding PKI, Second Edition.* Addison Wesley.

[9] Russ Housley, T. P. (2001) *Planning for PKI - Best Practices Guide for Deploying Public Key Infrastructure.* Wiley.

[10] Housley, R., Ford, W., Polk, W., and Solo, D., *Internet X.509 Public Key Infrastructure Certificate and Certificate Revocation List (CRL) Profile.* http://www.ietf.org/rfc/rfc3280.txt.

[11] Giessmann, E.-G. and Schmitz, R. (2000) *Zum Gültigkeitsmodell für elektronische Signaturen nach SigG und X.509. DUD,* **7**, 401–404.

[12] *Bildung von Stammzahl und bereichsspezifischem Personenkennzeichen (bPK).* http://www.cio.gv.at/it-infrastructure/sz-bpk/Stammzahl-bPKAlgorithmen.1-0-2.20040603.pdf.

[13] Schneier, B. (1996) *Applied Cryptography, Second Edition,* p. 470 ff. Wiley.

[14] Wolfgang Rankl, W. E. (2002) *Handbuch der Chipkarten: Aufbau - Funktionsweise - Einsatz von Smart Cards.* Hanser.

[15] Wang, X., Yin, Y. L., and Yu, H. (2005) *Finding collisions in the full sha-1. Lecture Notes in Computer Science,* **3621**, 17–36.

[16] *Certification Practice Statement für qualifizierte Zertifikate a.sign Premium.* http://www.signatur.rtr.at/repository.

[17] *Certification Practice Statement (CPS) Verwaltungs- und gewöhnliche Signatur Zertifikat.* http://www.signatur.rtr.at/repository.

[18] *PKCS #10: Public Key Cryptography Standard - Certification Request Syntax Standard.* ftp://ftp.rsasecurity.com/pub/pkcs/pkcs-10/pkcs-10v1_7.pdf.

[19] *Security Layer Spezifikation Version 1.2.0.* http://www.buergerkarte.at/konzept/securitylayer/spezifikation/aktuell/.

[20] *Microsoft Knowledge Base, How to turn off e-mail matching for certificates in Outlook.* http://support.microsoft.com/kb/276597/en-us.

Personal Firewalls for Linux Desktops

Andreas Gaupmann

SUSE Linux Products GmbH, Maxfeldstraße 5, 90409 Nürnberg, Deutschland

ABSTRACT

The usage of Linux as desktop system has increased steadily over the last few years. Therefore, Linux desktops have become an attractive target for attackers.

The traditional access control mechanism of Linux is discretionary. Users are allowed to modify access rights to their objects (e.g. files) at their own discretion. Effectively, such access control is determined by checking user identities. Such an access control mechanism does not suffice in a desktop usage scenario. Attackers may obtain user privileges by exploiting vulnerabilities of desktop applications. Accordingly, they gain access to every object belonging to the user.

Mandatory access control (MAC) was designed to eliminate the deficiencies of discretionary access control by imposing a set of rules onto a system. Access rights of users to objects are derived by a central authority. As a disadvantage, the process of administering and enforcing MAC policies is complicated and tedious for a user.

Personal firewalls deploy a user-friendly access control model that is based on events. Decisions on allowing or denying security events (application starts, incoming and outgoing network traffic) are taken directly by the user in an ad-hoc manner. A security policy is built up step by step as the events occur. Personal firewalls for Linux desktops are possible when several requirements are met: support by the kernel, modular design to facilitate privilege separation, and access control in user space.

Contact: andreas.gaupmann@fh-hagenberg.at
Availability:
http://developer.novell.com/wiki/index.php/Avalon

1 OVERVIEW

In this paper, the feasibility and requirements of a personal firewall for Linux are analyzed and the realized implementation is described. First, an introduction to the desktop threat model is given. Subsequently, requirements for personal firewalls on Linux systems are derived. Furthermore, an exemplary implementation is discussed. Finally, results and conclusions of the presented work are layed out.

2 PROTECTING DESKTOP USERS

Desktop systems are used by home and office users without a deeper knowledge of computers and networks. Web browsers, email clients, instant messengers, multimedia streaming clients, and VoIP softphones are typical applications that are used on desktop systems. All these programs interact with remote hosts in the Internet and process data received from them. Therefore, such applications are exposed to remote attacks that aim to compromise a desktop system or disclose private data of the user.

The weak spot of desktop systems are vulnerable applications. Exploiting applications is possible when errors have been made in one of the three big stages in the life cycle of an application (architecture and design, implementation, and operation). Three classes of threats for desktops may be distinguished.

- Security flaws in applications: Common programming errors facilitate buffer overflows, format string attacks, race conditions, integer overflows, and cross site scripting [1]. Insufficient or lacking input validation is the main cause for these flaws. Moreover, logical (semantic) programming errors may be used to divert the execution path of an application.

- Deception of the user: Phishing and pharming attacks redirect web browsers to servers that simulate popular websites (e.g. for providing online banking).
- Malware: Up to now, there have been no dangerous viruses, worms, or trojan horses that were targeted at Linux. Nevertheless, future malware (e.g. with multi–platform and polymorphic traits) may also attack Linux systems.

In the last decade, a multitude of security enhancements for Linux have been proposed. Some protect weak points of the operating system like process memory (PaX [2]) or filtering of network traffic (Netfilter [3]) while others (e.g. access control frameworks) confine vulnerable applications in a sandbox (AppArmor [4]) or control the entire system (SELinux [5]).

Access control frameworks differ in their approach on who is defining the rules for accessing resources. With descretionary access control (DAC) systems, a user can decide upon who may access her files by her own discretion. Mandatory Access Control (MAC) systems adhere to a rule set for accessing system resources which is centrally defined by a system security policy administrator.

DAC constitutes the traditional security policy on Unix–like systems. It is a useful model for home desktops because its user is typically also the administrator. Conversely, in a corporate environment desktop users with administrative privileges are an exception. In such a scenario, a personal firewall may use a predefined and fixed set of rules.

Typically, personal firewalls let the user build a rule set for determining access to resources. Therefore, personal firewalls combine DAC and MAC features.

3 PERSONAL FIREWALLS

The security policy that is implemented by personal firewalls can be described as event based access control. A security event is an action conducted by an application that may lead to the compromising of a system or the disclosure of private information. Examples are actions like starting binaries or communicating with hosts in the Internet. Four basic security events may be distinguished that have to be filtered in order to prevent, or at least hamper, attacks in the desktop threat model that was discussed in the previous section.

- Application starts: A program executes another program.
- Application replacements: An executable is replaced with a modified version of the file that contains additional (malicious) functionality.
- Outgoing connections: A program sends packets into the network.
- Incoming connections: A program receives packets from the network.

A security event may be caused by the logic of the concerned program or its successful exploitation. The personal firewall delegates the decision on denying or allowing a security event to the user. By this means, the user is given the possibility to control applications. For example, the execution of a shell by a web browser would be highly suspicious and is typically unwanted by the user.

Personal firewalls consist of three logical layers: the per–packet filter, the per–process filter, and the graphical user interface (GUI).

The per–packet filter is located between the network interface card (NIC) driver and the implementation of the TCP/IP stack. At this level, packets can only be filtered according to criteria taken from header or payload of the packet. Packets cannot be correlated with the application they are coming from (outgoing packets) or going to (incoming packets).

Per–process filtering takes place between the application and data transport layers. Generally, this is achieved from kernel space by wrapping the functionality of the transport driver interface. As a result, the functions for sending and receiving data into the network are intercepted. Socket operations are an example of such functions.

A personal firewall uses a GUI to interact with a user. It visualizes security events and displays logged information about previous user decisions. Additionally, the GUI allows the user to create, modify, and delete filter rules.

Personal firewalls provide little or no security if the attacker already has root privileges on the machine. The easiest thing the attacker can do is to disable the personal firewall. Probably, this will be recognized by the user. A more subtle method is to circumvent either the per–process or per–packet filter. This can be achieved by generating packets on a level below the particular filter by injecting them directly into the transport layer or the NIC driver, respectively.

The requirements for a personal firewall have been defined by the BSI [6]. The most important requirements are stated below. The design of the implemented personal firewall adheres to these criteria.

- Outgoing and incoming connections are filtered.
- Application starts are filtered in order to prevent the execution of untrusted programs.
- Security events are denied by default.
- Pre–defined rules are provided to allow common classes of network traffic by default (e.g. web surfing, instant messaging, file sharing, multimedia streams, local services, ...).
- Filtering is carried out per network interface to allow the definition of trust levels.
- The filtering actions are logged.
- A modular design facilitates extensibility.

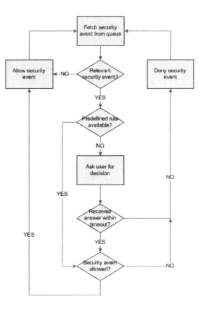

Fig. 1. The handling of security events by the decision layer.

4 REALIZATION OF THE PERSONAL FIREWALL

4.1 Design

The personal firewall takes advantage of a layered architecture by subdividing its functionality into components with differing privilege levels. The enforcement of filtering operations requires higher privileges than processing filtering rules or interacting with the user.

- Integration and enforcement: Decisions on allowing or denying security events can be enforced by hooking operations inside the kernel or creating an additional layer between kernel and user space. An attacker can circumvent the mediation layer in user space by directly calling kernel functions. As a result of this drawback, the discussed personal firewall for Linux desktops enforces rules in kernel space with a loadable kernel module.

- Decision making: The core component of a personal firewall is the decision layer. A major difference to common packet filters is the dynamic rule generation by directly asking the user. The findings from the analysis of a security event are evaluated and compared against a set of rules. If the security event is categorized as relevant, then it is processed and is either permitted or denied. Figure 1 visualizes this process in a flow chart. If a security event occurs while waiting on a user decision, then it is queued in a FIFO and handled as soon as the user has made a decision or a timeout is reached. In the latter case, the currently processed security event is denied by default.

- Storage of rules: Filter rules are stored in a database. This approach has been chosen because flexibility allows better extensibility. The analysis of complex rule sets requires a method for correlating and querying rule data. If a database is used for storing rules and logs, then the structured querying language (SQL) can be applied.

The personal firewall allows several classes of network traffic by default in order to increase the user experience. These network protocols include IGMP, ICMP, DHCP, and DNS.

4.2 Implementation

4.2.1 Kernel module The extensible Linux Security Modules (LSM) framework has been used for the implementation of the integration and enforcement layer. The LSM framework [7] has been introduced as standard part in the 2.6 line of the kernel to facilitate flexible access control [8]. A security module must be loaded into the LSM framework for implementing a specific security policy (e.g. event based access control).

The LSM framework adds an opaque security field to the structures of kernel objects along with functions to store data into this field and to retrieve it again. Furthermore, so called hook functions have been introduced in security critical portions of the kernel code. Every time such a code segment is traversed, the incorporated hook function will be called and its return value decides whether the requested operation is allowed or denied.

The LSM framework provides hook functions to control socket operations and program execution. Netfilter represents a well–tested packet filter framework as the widespread use of iptables for building secure Linux firewalls [9] shows. Therefore, these two kernel frameworks have been chosen as a basis for the kernel module of the personal firewall for Linux desktops. The hooks that are used by the personal firewall are listed subsequently.

- bprm_check_security: LSM hook used for the filtering of application starts.
- socket_connect: LSM hook used for the filtering of outgoing connections.
- socket_sendmsg: LSM hook used for the filtering of outgoing datagrams.
- NF_IP6_LOCAL_IN: Netfilter hook used for the filtering of incoming IPv6 traffic.
- NF_IP_LOCAL_IN: Netfilter hook used for the filtering of incoming IPv4 traffic.

4.2.2 Root daemon The root daemon utilizes one thread for reading messages from the kernel

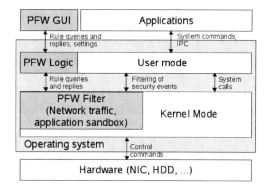

Fig. 2. The architecture of the implemented personal firewall.

module over a Netlink [10] socket and storing them in a message queue. Another thread reads the saved messages from the queue and reacts to them. The access to the message queue by the threads is synchronized with mutexes. Furthermore, the conditions of an empty or full message queue are checked and message processing is suspended until the situation is resolved. The additional step of storing messages in a queue is necessary because of the limited buffer size of a Netlink socket. The loss of messages can be avoided by keeping the buffer from filling up.

4.2.3 GUI The main tool of the user for interacting with the personal firewall represents the control panel shown in Figure 3. Notifications about security events are sent by the root daemon and received by the GUI over a Unix domain socket. As soon as a message is received, it is processed and displayed in the control panel. Once the user clicks on "Allow" or "Deny", the verdict is propagated down to the module and the row is removed from the table. The row is not removed if the user does not take a decision. In this case the timeout in the module will be reached and the security event will be denied. If the user renders a decision after this timeout, then the delayed answer message can still modify the filtering rules if the user requested it by activating the concerning checkbox.

A rule editor visualizes the rules for three classes of security events. Furthermore, the user can manage the set of rules of the personal firewall. Applied rules are listed in a history with information on date, time, and an indicator whether the rule has been stored persistently.

5 RESULTS AND CONCLUSIONS

It has been shown that the implementation of a personal firewall for Linux desktops is possible. Although the concept of a personal firewall is nothing new, the outlined work represents the first effort to bring about a user–friendly personal firewall for Linux.

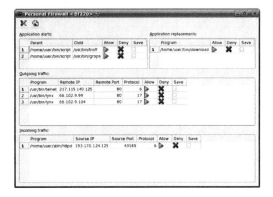

Fig. 3. The control panel organizes security events into tables. A table contains security events of the same class.

The findings show that the protection of a desktop system is possible by filtering certain classes of security events. Application starts, application replacements, incoming traffic, and outgoing traffic have been identified as security events that have to be filtered by a personal firewall.

Implementing a personal firewall functionality for a Linux system has proven to be difficult. The crucial questions have been how to enforce rules and how to get decisions on security events from the user space into the kernel space. Recent kernels provide solutions for both problems in the form of security frameworks (Linux Security Modules, Netfilter) and a communication infrastructure (Netlink). The security frameworks of the kernel, like the kernel itself, are subjects to constant change. Therefore, the presented implementation may have to adjust to these changes in the future. In fact, every security module that builds on these frameworks would have to be adapted. As a consequence, it is important to introduce a permanent interface for enforcing security policies into the Linux kernel. This task can be considered as an area of future research.

Finally, it is important to keep in mind that personal firewalls are building blocks within an overall security architecture. Moreover, with the predicted advent of cross–platform malware and an increased number of Linux desktops, personal firewalls are able to protect exactly the weak spot (application vulnerabilities) of these systems from being exploited.

6 REFERENCES

[1] Klein, T. (2004) *Buffer Overflows und Format-String-Schwachstellen.* dpunkt-Verlag.

[2] The PaX Team (2006), *Homepage of the pax team.* URL, pax.grsecurity.net/.

[3] Russel, R. (2006), *netfilter: Firewalling, nat, and packet mangling for linux.* URL, www.netfilter.org/.

[4] Cowan, C. (2006), *Novell apparmor: Application security for linux.* URL, www.fosdem.org/2006/index/speakers/slides.

[5] Smalley, S. and Morris, J. (2006), *Security-enhanced linux.* URL, www.nsa.gov/selinux/.

[6] Bundesamt für Sicherheit in der Informationstechnik (2006), *Anforderungen an module von sicherheitsgateways/firewalls.* URL, www.bsi.de/fachthem/sinet/fw-anf.htm.

[7] Wright, C., Kroah-Hartmann, G., Morris, J., Hallyn, S., and Smalley, S. (2006), *Linux security modules.* URL, lsm.immunix.org/.

[8] Wright, C., Cowan, C., Morris, J., Smalley, S., and Kroah-Hartman, G. (2002), *Linux security modules: General security support for the linux kernel.* URL, citeseer.ist.psu.edu/wright02linux.html.

[9] Barth, W. (2003) *Das Firewall Buch - Grundlagen, Aufbau und Betrieb sicherer Netzwerke mit Linux.* SuSE Press, 2rd edn.

[10] Salim, J., Khosravi, H., Kleen, A., and Kuznetsov, A. (2003), *Rfc 3549 - linux netlink as an ip services protocol.* URL, ietfreport.isoc.org/rfc/rfc3549.txt.

An Approach To NFC's Mode Switch[*]

Oliver Dillinger[a], Gerald Madlmayr[a], Christoph Schaffer[b], Josef Langer[c]

[a] Upper Austria University of Applied Sciences—Research and Development Competence Center, Hauptstraße 117, A-4232 Hagenberg, AUSTRIA
[b] Upper Austria University of Applied Sciences—Mobile Computing, Hauptstraße 117, A-4242 Hagenberg, AUSTRIA
[c] Upper Austria University of Applied Sciences—Hardware/Software Systems Engineering, Hauptstraße 117, A-4242 Hagenberg, AUSTRIA
http://www.nfc-research.at

ABSTRACT

NFC is an emerging technology for transmitting data over short distances. Its most basic component is the Mode Switch, which detects other devices in reach and propagates the device's own capabilities.

As this part of the firmware will be always running there are special requirements: low power consumption to ensure a long uptime and fast responses to satisfy the user. Besides it has to handle the difficulties of supporting the many different standards to which NFC maintains compatibility.

This paper analyses these requirements in detail and presents an implementation how to fulfill them.

Contact: oliver.dillinger@fh-hagenberg.at

1 INTRODUCTION

Near Field Communication[1] is a new communication technique for exchanging data over a short distance of about 10 centimeters. The base frequency is 13.76 MHz. Data encoding is based on existing standards for contactless smart cards. This should maintain compatibility with existing card and reader infrastructures. Supported standards are ISO 14443-A and B, and Felica (Sony's proprietary technology). On top of those RF layers, any other protocol layer can be used. Examples are ISO 14443-4, ISO7816, and EMV (Europay, Mastercard, Visa; a standard for credit/debit cards).

NFC itself is defined in ISO 18092 and uses ISO 14443-A's data encoding scheme for transfer rates of 106 kBit/s and Felica for higher speeds (currently, rates of 212 kBit/s and 424 kBit/s are defined, but may go up to 3 MBit/s in the future). NFC devices are able to interact with existing RFID readers by emulating card devices. To read contactless smart cards or RFID tags, it can act as a reader or writer, respectively.

It is also possible to exchange data between two NFC devices. This mode is called *peer-to-peer mode*. Similar to a card/reader combination one device acts as a master (the *initiator*), the other one as a slave (the *target*). Communication is always controlled by the initiator. The target only sends replies. Two modes are possbible: Either only the target emits a field for data transmission and the target answers by load modulating[2] this field, or both devices create a field alternately.

Like contactless smart cards, the target can be powered by the initiator's electromagnetic field. This could be useful if the target's battery was low.

---

[*] This project is sponsored by FFG (Austrian Research Promotion Agency).
[1] NFC

[2] A switched resistor is used to weaken the host's electromagnetic field.

2 REFERENCE HARDWARE DESIGN

As NFC is an emerging technology, nearly no devices exist which support it. For that reason we decided to develop our own reference board. Battery-powered, with a small form factor and a Bluetooth connection to the host, it is perfectly suited for mobile applications.

The printed circuit board has the size of a business card (see figure 1). Its main hardware components are:

- Analog circuitry
- Secure element
- Microcontroller
- Bluetooth connection

Enclosed in a plastic case, which houses power plug and switch, status LEDs and a rechargeable battery, the device has a thickness of less than two centimeters.

2.1 Analog Circuitry

The analog circuitry consists of an antenna with the necessary electric network for tuning and an NFC IC. This integrated circuit—a *Philips PN512*—provides all analog parts necessary. This includes modulation/demodulation of the data signal, detection of an incoming signal's modulation type and speed (necessary when in target mode) and dealing with some low-level protocol issues.

The NFC IC supports two data sources: a microcontroller (2.3), which might emulate cards, act as a card reader, or do communication over NFC, and a secure element (2.2).

As the antenna is a very critical component in terms of coverage and bit error rate, its design has been copied from an implementation whitepaper.

2.2 Secure Element

The secure element is a microcontroller equipped with a crypto core and is connected directly to the NFC frontend. It runs a dedicated operating system (Java Card Open Platform). Through its architecture it is possible to securely store data (e. g. private keys, passwords). Applets loaded into the chip run in sandboxes and can't interfere. Additional security is given by a special chip design which makes it hard to infer the applet's code by observing the current consumption; it is also difficult to extract stored data by opening the chip, for example.

In our reference implementation we use a *Philips SmartMX* secure element. This device is not necessarily a separate component. In mobile phones, for example, it may be included in the SIM or a secure SD card.

Though a secure element seems to be vital for many applications (ticketing [1], e-purses) NFC's standardization body NFC-Forum[3] decided that it is not a mandatory part in NFC devices.

2.3 Microcontroller

An *Atmel ATmega128*[4] is used to run the firmware. It handles communication with the host, controls the NFC frontend and deals with NFC's protocol issues.

2.4 Bluetooth Connection

The Bluetooth Module provides a serial link to the host using BT RFCOMM protocol. All necessary software for the Bluetooth stack is integrated within the module.

Bluetooth was chosen because of its widespread use in the mobile world. Almost all modern cell phones and PDAs have a Bluetooth interface. Thus, the reference board is of great use when doing NFC application with todays mobile devices.

3 MODE SWITCHING

The Mode Switch is responsible for detecting other NFC devices within reach and advertising the own device's presence and capabilities to others. As this is the basic, always running entry point to the protocol stack, special care must be taken with its design.

---

[3] http://www.nfc-forum.org
[4] http://www.atmel.com

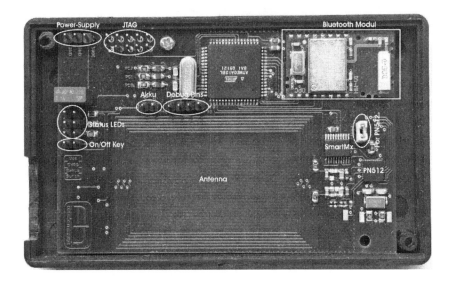

Fig. 1. PCB

3.1 Principle

In classic RFID infrastructures the roles were clear: readers would always be active and cards would always be passive (i.e. cards are never initializing communication). A situation where two readers were so close that their electromagnetic fields would interfere was nearly impossible. In such a case the system simply wouldn't work.

With NFC the conditions are different. As every device can initiate communication, a method is needed to avoid chaos if several devices are close together. This method is called *Mode Switch*. It has its name from the fact that the device continuously switches between target mode (acts like a card, listens) and initiator mode (acts like a reader, polls). Because NFC must support several different contactless smart card standards, the polling phase of the mode switch consists of several request commands (see 3.2.4).

Figure 2 shows the principle.

3.2 Analysis of Requirements

The Mode Switch is the component of the protocol stack that will always be active and serves as entry gate.

Thus, there are several requirements it should meet:

- Low power consumption
- Fast detection of other devices
- Prevention of collisions
- Support of all required contactless smart card standards

3.2.1 Power Consumption Most NFC-enabled devices will be battery-powered. To ensure a long uptime—which customers expect—it is necessary that Mode Switching uses as little energy as possible.

Certainly, most power is needed when the device acts as a reader, because it has to generate an electromagnetic field. Therefore, the polling interval should be as long as feasible (see also next point), with the field being kept up for the shortest time possible.

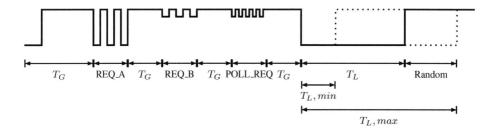

Fig. 2. Mode Switch principle

Another way to conserve energy would be to use *passive mode communication*. In this mode, only the initiator creates a field, while the target modulates it to send data. Contrary, in *active mode communication* both target and initiator alternately generate a field.

3.2.2 Fast Detection A good user experience requires that devices coming into reach must be detected as fast as possible. Studies show that a delay of more than 200 milliseconds is considered being "too long" by test subjects.

Of course, the goal of fast detection which requires much polling, and the goal to conserve energy are contrary. A trade-off between both has to be made, bearing a satisfactory user experience while consuming not too much battery lifetime.

3.2.3 Collision Prevention Every NFC device may act as an initiator. If two devices initiate a field and send data at the same time, a *collision* occurs. To detect them, a mechanism called *Initial RF collision avoidance (RFCA)* is employed [2].

RFCA ensures that there is no external field for a certain amount of time before the device tries to generate one on its own. This delay is calculated using the following formula:

$$T_{RFCA} = T_{IDT} + n \times T_{RFW} \quad (1)$$

with

Standard	Example
ISO14443-A	Philips Mifare, NFC
ISO14443-B	Digital Passport
Felica	Sony proprietary, NFC
Others	t. b. d.

Table 1. Supported standards

T_{IDT}	Initial delay time
T_{RFW}	RF waiting time
n	Randomly generated, between 0 and 3

3.2.4 Supported Standards As NFC should maintain compatibility with existing contactless smartcard infrastructure it has to support many different standards. Although an NFC device is not required to support card emulation or Reader/Writer emulation, the standards listed in table 1 are discussed to be part of NFC for downwards compatible devices.

This means that the mode switch not only has to detect other devices but also find out of which technology they are. This means that one polling cycle consists of many different request commands, often also with different RF specifications (i. e. different modulation indizes or the like).

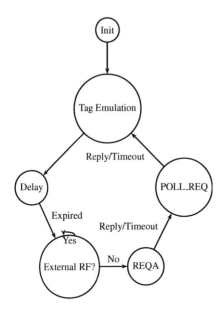

Fig. 3. Firmware flow chart

3.3 Implementation

Firmware is implemented in ANSI C and compiled with GCC for AVR target[5]. It is an approach of a firmware implementation that tries to fulfill the requirements set up in the previous section.

Figure 3 shows the firmware's structure. An infinite main loop executes target mode first. Now the device can be discovered and used by other initiators (it doesn't matter whether it's NFC or legacy). It keeps this state for a configurable amount of time, usally below 200 milliseconds (see 3.2.2). A random element may be added to avoid synchronosity between devices. After the time expired, the firmware checks for the presence of an external field and doesn't leave tag emulation until there is no field present. This helps avoid problems with legacy infrastructure: without this check, there would be small pauses in card emulation when the device tries to switch

to initiator mode and executes RFCA. If a legacy reader wants to access the fake card right then, it would be unable to find it.

While in card emulation, other initiators actually will detect two devices: NFC (NFCID1 and NFCID2, respectively), and the secure element. The peer is completely free to decide which ID to select.

The next step is starting initiator mode. After doing a successful RFCA an RF field is generated and request commands are sent. Right now two standards are supported: ISO 14443-A (*Type A*) and Felica. Other standards are currently being discussed in the NFC Forum.

When polling for different technologies, the order is of importance, because it is not clear how cards will deal with foreign commands. For example, Type A-cards use an ASK modulation index of 100%, compared to 10% with ISO 14443-B (*Type B*)-cards [3]. It is possible that a Type B-card fed with Type A commands resets and loses its state due to power shortages caused by the excessive modulation of the field.

Another issue when looking for NFC-enabled devices is the fact that they use two different protocols for data transmission: ISO 14443-A and Felica. It bears a different ID in both standards. If an NFC device is polled with a REQA command, its internal mode selector will adapt to the Type A radio interface and do not respond to a subsequent Felica POLL_REQ. Thus, if a break in the RF field occurs, the mode selector will reset and the NFC device will answer to a POLL_REQ. This leads to the (wrong) detection of two devices that can't be matched because of the different IDs.

The system's power consumption is minimized by putting the microcontroller into sleep mode while waiting for events (data reception, timeouts) and waking it up through interrupts. Additionally, only the registers necessary are changed in the NFC frontend IC. Another approach would have been to rewrite all configuration registers when switching to a different mode, but that would need both more time and energy.

---

[5] http://winavr.sourceforge.net

3.4 Compatibility With Existing Infrastructure

What's necessary for maintaining compatibility with existing systems is heavily discussed in the standardization group. We have tested our device with three systems, two legacy and one NFC.

3.4.1 Philips Pegoda The *Philips Pegoda* is a reader for contactless smart cards. When using in normal mode with constant field, our NFC box is recognized and works normally as a fake card. When using a polling mode with an intermittent field, things are different: no relaible detection.

This seems to be a principal problem with polling legacy readers that use no RFCA. When they switch off their field the NFC device may switch to initiator mode. As the reader doesn't expect another reader in reach, it just switches its field on again, producing a collision. Dependent on the delay times of the two applicances, the result is an unsteady detection, i. e. some polling cycles will succeed, others will fail.

We expect that this problem can't be solved without proper software updates for existing systems.

3.4.2 Mifare Access Control Reader Readers that can get a proximity card's Mifare ID. Their initial firmware version did not emit a constant field for power saving reasons, which led to problems similar to 3.4.1. After a software update that led to a constant field, our NFC device stays in card emulation permanently and can be detected.

3.4.3 Samsung SGH-X700n One of the first available mobile phones with NFC support. Due to issues in its firmware, it needs a time of about 400 milliseconds after a polling request to settle again. If this timing constraint is not met, some subsequent requests will fail. Otherwise, detection is satisfactory.

4 CONCLUSION

As a basic component, NFC's Mode Switch needs to be both power efficient and fast. Our proposal includes an analysis of these requirements and a reference implementation of how they could be reached. Testing the algorithm showed that there may be some principal problems with the compatibility with existing infrastructure.

5 REFERENCES

[1] Madlmayr, G., Dillinger, O., Gierlinger, S., Kleebauer, P., Pucher, R., Vymazal, D., Langer, J., and Schaffer, C. (2006), Secure Token Concept within a Near Field Communication Ecosystem.

[2] ISO (2004), Near Field Communication - Interface and Protocol (NFCIP-1), ISO/IEC 18092.

[3] ISO (1999), Identification Cards - Part 2: Radio frequency power and signal interface, ISO/IEC 14443-2.

Model-Driven Development of Speech-Enabled Applications

Werner Kurschl[a,*], Stefan Mitsch[b], Rene Prokop[b], Johannes Schönböck[b]

[a] Upper Austria University of Applied Sciences—College of Information Technology, Hauptstraße 117, A-4232 Hagenberg, AUSTRIA,
[b] Upper Austria University of Applied Sciences—Research and Development Competence Center, Hauptstraße 117, A-4232 Hagenberg, AUSTRIA

ABSTRACT

Motivation: Interacting with a computer by speech—like humans do among each other—is a dream software engineers work on to come true since the 1960s. When using devices that lack adequate input capabilities, like personal digital assistants (PDA) or mobile phones, everybody would prefer a more convenient interaction—speech. But we see two major hindering factors: the processing power of these devices is not yet sufficient and speech recognition engines lack application developer support.

Results: We propose a highly configurable software architecture that divides speech recognition in processing steps that can be distributed among several devices and, thus, allows speech recognition also on limited devices. Additionally, we use Model Driven Development to unify graphical and voice user interface development. Thus, graphical and voice user interfaces need not be developed separately, but are generated from a single model. To lower the entry barrier for application developers, the user interface framework bases on the same paradigms and components (button, textfield, etc.) as graphical user interface frameworks do.

Availability: July 2007
Contact: werner.kurschl@fh-hagenberg.at

[*] to whom correspondence should be addressed

1 INTRODUCTION

Technological developments in mobile devices (e.g., ubiquitous network access) through the past years enabled developers to implement collaborative business applications that allow traveling field workers to cooperate with workers at stationary desktop PCs. Software is in such an application scenario used as a compensation for direct collaboration between coworkers; hence, there is the desire for interacting with devices through natural ways.

Interacting with a computer by speech—like humans do among each other—is a dream software engineers work on to come true since the 1960s. Today's speech recognition systems for desktop PCs reach an impressive accuracy of about 95%. This has largely been the result of increasing processing power combined with improvement of algorithms. Yet, speech recognition has not been adopted by a broad range of users; it remained a niche product for mainly the medical and legal profession. One reason for this could be that users prefer to stick to conventional input methods like keyboard and mouse, rather than learn to handle new input devices. Medics and lawyers are used to dictating and can benefit from speech recognition without changing their habits. But when using devices that lack adequate input capabilities, like personal digital assistants (PDA) or mobile phones do, everybody would prefer a more convenient interaction—speech. Unfortunately, the processing power of these devices is not yet sufficient for unlimited speech processing.

Speech recognition systems can roughly be categorized by the following criteria: (a) Speaker-dependent or speaker-independent (b) constrained (grammar-based) or unconstrained (dictation-based) vocabulary. Today's speech recognition systems that allow unconstrained vocabulary are speaker-dependent and demand high processing power and memory and are therefore only available on desktop PCs. In contrast, constrained vocabulary speech recognition systems are often speaker-independent and available on mobile devices as well.

Another hindering factor for the success of speech recognition is the support for application developers. While graphical user interfaces can be developed using tools and editors based on well-established and approved paradigms and components, voice user interfaces currently require application developers to go back to the roots of user interface programming, which increases the costs of developing speech-enabled applications.

The diversity of devices utilized in collaborative environments is a challenging factor in implementing software systems. Often, applications created for a desktop PC are transformed to applications suitable for mobile devices. This transformation, when done manually, is tedious, error-prone, and increases the cost of software development. Thus, there is a strong demand to create an application once at a more abstract level, and transform this abstract model automatically into applications suitable to various devices.

2 RELATED WORK

VoiceXML (see [1]) is a markup language add-on to HTML pages that is interpreted by special browsers. It enables voice-controlled Web pages, but has two disadvantages: it allows only constrained speech recognition, which limits the application domain, and it needs a continuous connection to a Web server that hosts VoiceXML sites.

XUL (see [2]) and UIML (see [3] and [4]) are both user interface markup languages based on XML. User interfaces can be described in XML format and can then be transformed into multimodal, VoiceXML based user interfaces.

Multimodal Teresa (see [5]) offers a model-based approach to design (speech-enabled) applications for different platforms. An abstract task model is the basis for generating a platform specific model (PSM). The PSM can then be used to generate an application's user interface implementation. Although Teresa offers an editor, the programmer has little influence on the final look and feel that is generated through transformations. Furthermore, voice user interfaces and multimodal interfaces are expressed with VoiceXML, which restricts the approach to Web applications.

The MONA (see [6]) research group developed a multimodal presentation server. The basic idea is to generate a device-specific implementation of a user interface based on an abstract description. The format is a MONA-specific UIML format that is transported to the mobile device via Web services. The mobile device must be capable of interpreting this user interface description in a specific browser. If a multimodal user interface is needed, then VoiceXML is used as target language.

Microsoft's Speech API provides a common API to deal with different recognition engines. Microsoft intends to implement the SAPI for mobile devices but to date, there is no implementation available.

Aurora Distributed Speech Recognition (DSR) (see [7]) is a standard of the ETSI. DSR enables speech recognition on a remote machine by sending speech data over low bandwidth networks. Therefore, speech is not transmitted as is, but instead so-called features are extracted from speech data. On the server side a speech recognition engine, which must be able to process the extracted features, converts them into text. Although DSR is standardized, hardly any dictation engine is able to recognize speech from features.

We identified the following shortcomings in the described research work, when applied to our

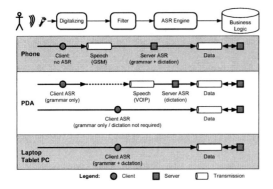

Fig. 1. Speech recognition scenarios.

application domain. Creating distributed speech applications with the Aurora standard limits the user to special speech recognition engines, which hinders the wider adoption of speech technology in business applications. Most related work in the field of automatic model transformation from user interface descriptions to user interfaces suggests VoiceXML as language for multimodal applications, as it is simple to transform to (its XML after all) and allows one to develop Web applications with a voice user interface easily. But it is not easily possible to create stand-alone applications based on VoiceXML. Moreover, our analysis showed that, just as all other examined frameworks, it does not support dictating free text; all are restricted to constrained speech recognition. Hence, the current limitations of existing approaches do not allow the development of a seamless model and framework that fits different devices with varying capabilities.

3 MOBILE SPEECH RECOGNITION

Speech recognition consists of three basic steps (depicted in Figure 1): digitalizing, filtering, and automatic speech recognition.

The first step in speech recognition is recording and digitalizing spoken language. It results in a digital signal that represents what was spoken. This digital signal's quality is in the second step enhanced through filtering. Typical filters perform audio volume adjustment, noise reduction, or audio compression. Finally, in the third step automated speech recognition is performed. It transcribes the digital audio signal into text.

Each step consists of components (we refer to them as speech components hereafter) that, chained together, form a speech recognition system. The speech components can be distributed among several devices; thus, each device hosts a part of the speech recognition system it is capable of. The distribution leads to a set of implementation scenarios (also shown in Figure 1).

Scenario 1—Phone: Conventional (cell) phones are only capable of recording, digitalizing and transmitting speech over a GSM network. A remote device hosts the remaining speech components (filters and ASR engine).

Scenario 2—PDA or Smartphone: PDAs and Smartphones have, compared to conventional phones, enough processing power to host additional speech components (e.g., constrained speech recognition, but not dictation). When dictation is required, the speech recognition engine needs to be hosted on a remote machine, while some of the filters might still be hosted on the mobile device (e.g., to ensure higher audio quality, or to keep the bandwidth requirements low).

Scenario 3—Laptop or TabletPC: Portable PCs can host the entire speech recognition system.

4 ARCHITECTURE

A typical enterprise software system (shown in Figure 2) that is extended with a voice user interface consists of four main parts: (a) the client software, (b) a local speech recognition engine, (c) a remote speech recognition engine, and (d) a business server. We base our architecture on scenario 2 described in the previous section, in less generic scenarios some of these parts are optional.

The client software consists of a multimodal user interface and a business logic layer that communicates with an enterprise server; the enterprise server typically hosts business logic and

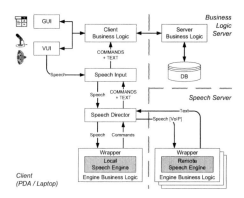

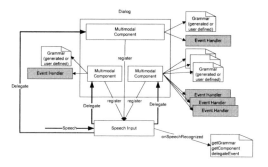

Fig. 2. Architectural overview.

Fig. 3. Speech input and event handling.

keeps persistent data in a database. An abstraction layer—the *Speech Input*—allows the client business logic layer to access the speech recognition system. It hides the deployment of the speech components; thus, the application can use a voice user interface completely independent from the type and processing power of the mobile device. The speech recognition system consists of a local speech engine that handles constrained command-and-control tasks, and of a remote speech engine that allows dictation. This ensures low latency for tasks where the user needs immediate feedback (e.g., executing a command), but also allows dictation on devices that are typically not capable of hosting a full-fledged speech engine. The *Speech Director*, which is an implementation of the Message Router (see [8]), controls the cooperation of the local and the remote speech engine. It routes speech data to the appropriate engine and merges the returned results. Hence, the number, type, and location of the engines are completely hidden from the client.

Speech components are linked through communication channels that transport speech data; together they form a Pipes-and-Filters architecture (see [9]). A channel can be implemented using various technologies, which leads to flexibility in deploying the speech components: simple streams can be used for transporting data between speech components hosted on a single device, while VoIP, GSM, or similar protocols can be used to transmit data between speech components hosted on different devices.

5 USER INTERFACE DEVELOPMENT

Today's frameworks for building graphical user interfaces typically use events to communicate user input to an application. This paradigm is well established and therefore familiar to most developers. Handling user input from a voice user interface should base on a comparable paradigm; developers want to handle speech input just as easy as they handle input from a mouse or keyboard. But current speech recognition engines provide a generic "speech recognized" event. Developers have to deal with the details of the event: often they have to examine the recognized text to decide which action to take.

Our framework provides multimodal components, as shown in Figure 3, that provide meaningful events (e.g., text changed, item selected) from both traditional and speech input. Registered event handlers define code that is executed if a certain event occurs. Each component defines its syntax (i.e., valid input values) in a grammar. Simple components, like voice-enabled buttons, provide a generated default-grammar, but they can also be customized with a developer-specified grammar.

Multimodal components are grouped in *dialogs* by the task they solve. When a dialog is activated, it registers its components at the Speech Input, which in turn activates the components' grammars in the speech recognition engines. Upon the engine's speech recognized event, the Speech Input translates the results into events.

The framework's implementation of the multimodal components was chosen from the following alternatives: (a) Custom Controls, (b) UIML, and (c) MDA. Each approach's consequences for application developers are discussed in detail in the following subsections.

5.1 Custom Controls

The .NET Framework and Java provide a simple extension mechanism for user interface components. These so-called custom controls can be integrated in the development environment and, thus, be used very conveniently. For example, a voice-enabled button would be a custom control class derived from Button or JButton; its inherited behavior is extended by a grammar and additional properties/methods that allow the Speech Input to issue events.

For mixed initiative dialogs, during which the user can provide several information items at once, additional components without graphical representation would be needed. These components might be difficult to use for developers; but more importantly, they lead to separate implementations for the graphical and voice part of a multimodal user interface. The main drawbacks of this approach for application developers are:

- Each platform needs its own controls that might differ from other platforms.
- Each programming language needs different controls.
- User interface development is restricted by the development environment (which is most often designed for building graphical user interfaces).
- New controls (e.g., for mixed initiative dialogs) are exposed to the developer.

5.2 User Interface Markup Language (UIML)

The User Interface Markup Language (UIML) is an XML-compliant domain specific language that allows developers to describe a graphical user interface and its layout independently from the platform and implementation language used. So-called rendering engines (one for each platform and implementation language) transform the description at runtime into user interface components. Hence, UIML does not suffer from the drawbacks mentioned in the previous section. But to date, UIML lacks support of a graphical editor; user interface descriptions need to be manually specified in XML format, which makes imagining the final outcome difficult. Moreover, there are hardly any rendering engines available and the support for multimodal user interface development is still under research. Most research work in this field is based on VoiceXML (see, e.g., [5], and [6]), which is not applicable in our domain for the above mentioned reasons (restricted to Web applications, no dictation, etc.).

5.3 Model Driven Development (MDD)

Due to the fact that there is no tool that completely satisfies our needs we introduce our own editor that allows model-driven software development. Model-driven development (MDD) raises the level of abstraction at which a software developer works; its main goals are simplifying and formalizing software development activities, so that automation is possible (see [10]). Thus, domain experts are able to create models at a high level of abstraction without programming knowledge.

As there is to date no common understanding of the terms and definitions used in MDD, we briefly specify the terms used subsequently (see [10] for a complete definition). A *model* is an abstraction of a software system consisting of elements. The elements of a model depend on the domain described by the model (a model describing the graphical user interface of a software system contains elements for Buttons, Labels, etc.). A software

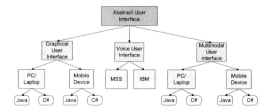

Fig. 4. Models with different levels of abstraction.

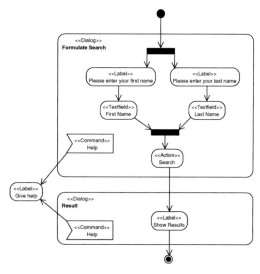

Fig. 5. A sample workflow modeled in UML.

system can be described through multiple complementary models that differ in their level of abstraction.

In OMG's specification of model-driven architecture (MDA, see [11]), the different levels of abstraction are called platform independent model (PIM) and platform specific model (PSM); the PIM is on a higher level of abstraction than the PSM is. The PSMs can be generated based on the PIM—this step is called *transformation*. To automate transformations between models, so-called *relationships* map elements of one model to elements of another model.

Figure 4 shows that in our approach an abstract user interface model is the basic PIM for a user interface. On this level, properties common to all types of user interfaces can be modeled. The next more concrete level of abstraction contains three possible models: a model for a *graphical-only user interface*, a model for a *voice-only user interface*, and a model for a *multimodal user interface*. These three models describe details inherent to the modeled type of user interface. Each of these models can then be further refined for specific platforms (e.g., PC or mobile device). The final (and most concrete) model is the implementation in a particular programming language.

Within the abstract user interface model it is possible to create the *workflow* of an application. This workflow defines the application life cycle, i.e., the interaction sequence of a user with the application. So-called *dialogs* are used to represent interactions that belong together, as shown in Figure 5. A common strategy in requirements engineering is Use Case Modelling, which captures an application's requirements. In UML, use cases can be further refined in activity diagrams, which describe the use case in terms of a sequence of actions. The refined use cases are an excellent basis for user interface design; hence, we use activity diagrams for describing our abstract user interface model.

A dialog consists of abstract user interface elements; currently, we defined *Action*, *Single Selection*, *Multiple Selection*, *Label*, *Textfield*, and *Command*. Actions represent user components that are used for confirmation or denial. Single Selection is used to select a single value from a fixed set of predefined values, while a Multiple Selection is used to select several values. Commands can be used to navigate through the application or to provide help to the user.

Table 1 shows the relationships between the abstract, the graphical, and the voice user interface model's elements; the relationships describe how the abstract user interface model can be transformed to a graphical or voice user interface model.

Table 1. Relationships between elements of different models

Abstract	Graphical	Voice
Dialog	Form	Dialog
Label	Label	Prompt
Textfield	Textfield	User Reply
Action	Button	Command
Command	Menu Entry	Command
Single Selection	Drow-Down-List	User Reply
Multiple Selection	Multi-Selection-List	User Reply

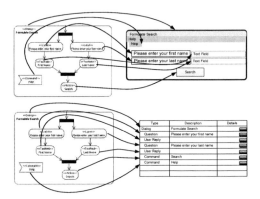

Fig. 6. Transformation from the abstract to the graphical and voice user interface model.

Abstract user interface elements are mapped to common graphical elements that we think need no further explanation. The voice user elements in the table have the following semantics: a *dialog* is a sequence of one or more speech activities that allow a contiguous communication between the application and the user. A *prompt* represents an application utterance without expecting feedback from the user. Input from the user is captured by a *user reply*. In the common scenario where prompts and user replies are used together, the prompt may also be titled *question*. A *command* leads to an immediate response of the application.

In the graphical user interface model the layout of the graphical elements is specified. In contrast to a graphical user interface model, in a voice-only user interface model the layout of the elements is unimportant; instead, the focus is on the verbal interaction between the application and the user. This is typically specified in the form of grammars that are attached to user interface elements. Note that a grammar may not only be added to a specific component but also to the dialog itself. In this case a grammar that allows the user to fill in all elements with a single sentence could be attached to the dialog.

As Figure 6 shows we try to provide a realistic preview of the user interface within the model driven editor. Using realistic elements leads to a better understanding of the application during design and development time. Additionally, these models can already be presented to users that can give accurate feedback on the design; we expect reduced development time and increased user satisfaction from this.

The proposed MDA editor offers the most flexible tool compared to custom controls and UIML. Due to the fact that all models are derived from the abstract user interface model a developer only has to create a model once. Transforming this model to different platforms and different devices can be done automatically. This helps to reduce errors and to reduce development time. The developer is not limited by the restrictions of a certain development environment. Compared to UIML we add a graphical representation of the domain specific language. This may reduce initial training time as well as development time of an application.

6 CONCLUSION

We present a component-based framework that supports the development of multimodal applications on heterogeneous devices ranging from laptop PCs to mobile phones. Our work focuses on distributable components that allow speech recognition on any device. The speech recognition process is split into small steps that can be distributed flexibly depending on a device's capabilities. The presented approach eases the development of speech-based applications. The framework is partially implemented and shows promising results in dictating text on a PDA.

To ease application development, we propose a model-driven development approach for developing user interfaces. We discussed different possibilities to create user interfaces for different platforms and various modalities (graphical, multimodal or voice only). The proposed MDA editor provides an abstract user interface model to define the basic workflow and the basic components of an application. It is the basis for generating more specific models and, finally, source code. This part of our work is currently under development; future work comprises the definition of an event model, the implementation of model transformation tools, and advanced editors.

7 ACKNOWLEDGEMENT

We thank Peter Schranz, Georg Schabetsberger and Heiko Rahmel for their support and valuable input. This research was supported by the Austrian Research Promotion Agency under the FHplus program, the Austrian Broadcasting Corporation (ORF) and Microsoft. Any opinions, findings, and conclusions or recommendations in this paper are those of the authors and do not necessarily represent the views of the research sponsors.

8 REFERENCES

[1] McGlashan, S., Burnett, D. C., Carter, J., Danielson, P., Ferrans, J., Hunt, A., Lucas, B., Porter, B., Rehor, K., and Tryphonas, S. (2004), Voice Extensible Markup Language (VoiceXML) Version 2.0, W3C Proposed Recommendation. http://www.w3.org/TR/voicexml20.

[2] Goodger, B., Hickson, I., Hyatt, D., and Waterson, C. (2001), XML User Interface Language (XUL) 1.0. http://www.mozilla.org/projects/xul/.

[3] Abrams, M. (2000), User Interface Markup Language (UIML) Draft Specification. http://www.uiml.org/specs/docs/uiml20-17jan00.pdf.

[4] Phanouriou, C. (2000) *UIML: An Appliance-Independent XML User Interface Language*. Ph.D. thesis, Virginia Polytechnic Institute and State University, Blacksburg, VA, USA.

[5] Paterno, F. and Santoro, C. (2002) One Model, Many Interfaces. *Proceedings of 4th International Conference on Computer-Aided Design of User Interfaces (CADUI)*, Valenciennes, France, pp. 143–154, Kluwer Academics.

[6] Niklfeld, G., Anegg, H., Pucher, M., Schatz, R., Simon, R., Wegscheider, F., Gassner, A., Jank, M., and Pospischil, G. (2005) Device Independent Mobile Multimodal User Interfaces with the MONA Multimodal Presentation Server. *Proceedings of Eurescom Summit 2005*, Heidelberg, Germany.

[7] Pearce, D. (2000) Enabling New Speech Driven Services for Mobile Devices: An Overview of the ETSI Standards Activities for Distributed Speech Recognition Frontends. *Proceedings of AVIOS 2000: The Speech Applications Conference*, San Jose, CA, USA.

[8] Hohpe, G. and Woolf, B. (2004) *Enterprise Integration Patterns - Designing, Building, and Deploying Messaging Solutions*. Addison-Wesley.

[9] Buschmann, F., Meunier, R., Rohnert, H., Sommerlad, P., and Stal, M. (1996) *Pattern-Oriented Software Architecture - A System of Patterns*. John Wiley & Sons.

[10] Hailpern, B. and Tarr, P. (2006) Model-driven Development: The Good, the Bad, and the Ugly. *IBM Systems Journal*, **45**, 451–461.

[11] Petrasch, R. and Meimberg, O. (2006) *Model Driven Architecture - Eine praxisorientierte Einführung in die MDA*. dpunkt.verlag.

Evaluation of Temperature-aware Quality of Service

Thomas Plachy

Interuniversitary Microelectronics Center (IMEC), Kapeldreef 75, B-3001 Leuven, Belgium

ABSTRACT

Moores law says that chip density doubles every 18 months as well as memory density and power consumption. This has been made possible by scaling down the technology node. Power is becoming the main design constraint in microprocessors. To be aware of the power dissipation is the most crucial thing for low power processors, because they are used for mobile devices and are powered by battery. In this work we implemented a multimedia application to explore the impact of power and temperature-aware techniques.

Motivation: For portable applications a low energy consumption is crucial. Downscaling to new architecture technologies reduces the dynamic power consumption. However, the leakage currents, the power density and the die temperature increase with downscaling. The energy cost and leakage power may exceed the energy savings from scaling. Hence, the semiconductor industry may hit a wall. Reducing leakage currents is therefore crucial. Leakage currents are particularly important in memory. Due to the fact that leakage currents are dependent on the temperature it is necessary to reduce the die temperature to lower the leakage currents. Another reason is that the cooling costs are rising with increasing power density. Processor packaging becomes the major expense and cheap packages can no longer be designed for worst case. So there is an urgent need to reduce or control heat dissipation. Run-time techniques that can regulate the temperature are necessary because they can prevent thermal emergencies by changing the processors activity. Thermal models are required to evaluate such techniques.

Results: Two techniques have been implemented: frequency scaling and Quality of Service (QoS). We illustrate that frequency scaling alone does not deliver enough efficieny. Timing violations and signal distortions may occur. So it is crucial to use a combination of Quality of Service method and frequency scaling which reduces signal accuracy, but avoids signal distortions and timing violations.

Contact: thomas.plachy@fh-hagenberg.at

1 INTRODUCTION

Since the introduction of microprocessors in the early 1970s their performance has increased exponentially. Moore's law says that chip density doubles every 18 months as well as memory size and power consumption. This problems have been solved by scaling down the technology node and now we count transistors inside microprocessors in billions and clock frequency in gigahertz. A lot of transistor technologies have been developed but only Complementary Metal Oxide Semiconductor (CMOS) became the most popular and commonly used technology. Transistor-Transistor Logic (TTL) and Emitter Coupled Logic (ECL) could not achieve against CMOS in the mainstream production and are only used for a very small spectrum of applications. The reason of the high popularity of CMOS is the low static power dissipation in circuits. Power is becoming the main design constraint in microprocessors, because the trend for portable battery-powered devices such as cellular phones, PDAs and portable computers is still growing. They are battery-powered and therefore it is important to reduce the power consumption to a minimum to guarantee a reasonable runtime.

2 GENERATION OF LEAKAGE CURRENTS

Dynamic power is the major source of total power consumption in today's microprocessors and is caused by charging and discharging CMOS circuits. Nevertheless, static power which is caused by leakage currents is getting more and more important. Scaling increases both kinds of power dissipation. The International Technology Roadmap for Semiconductors (ITRS) [4] predicts that in the next processor generations static power may exceed dynamic power (see figure 1).

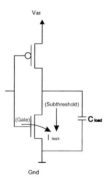

Fig. 2. Subthreshold current and gate-oxide current are the two main contributors of leakage current.

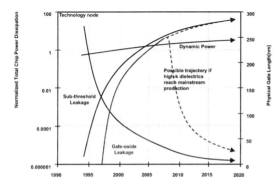

Fig. 1. Trends on static, dynamic and leakage power consumption based on the International Technology Road-map for Semiconductors (ITRS)2002.

The two main contributors of leakage currents are the subthreshold current and the gate-oxide current (see figure 2).

$$I_{leak} = I_{sub} + I_{ox} \qquad (1)$$

Subthreshold leakage current flows from the drain to the source. To scale the performance the threshold voltage of the transistor must be be reduced. Gate-oxide leakage current is running through the oxide layer. Scaling reduces the thickness of the oxide layer and increases the leakage [2].

A fundamental problem is that these currents are still flowing when transistors are switched off. Leakage currents specially occur in memories because they are built up on four or six transistors. In State-of-the-art microprocessor designs large parts of the chip are dedicated to memory structures [6], [7]. A lot of research was done to reduce leakage currents in memory modules like caches, registers and tag arrays. Leakage control at architecture level is attractive, because large groups of circuits can be controlled. Most commonly used techniques are dynamic voltage scaling [5], frequency scaling and multiple threshold transistors [8].

The problem is that most of these methods use only an abstract model of leakage and do not account all effects that may impact leakage currents, like supply voltage and temperature. Especially temperature has a major impact on leakage currents. Leakage currents increase with rising temperature. Chip temperature changes due to the activity and the ambient temperature. But also power dissipation affects the chip temperature. Variations of switching activity across the die and diversity of the type of logic results in uneven power dissipation across the die. This variation results in an uneven distribution of the supply voltage and temperature hot spots. Hot spots are degrading the reliability of the system and may cause thermal runaway and damage of the chip. A reduction of the temperature will positively influence the leakage and can be obtained by decreasing the clock frequency or reducing the supply voltage at runtime. At the design level, temperature can be reduced by using packages with a better heat conductivity and implementing a heat spreader inside the package for a better heat distribution.

3 LEAKAGE OF ARCHITECTURES

Many different types of processor architectures are available on the market. To keep track of all of them, chip designers categorize the processors in several different types (see figure 3).

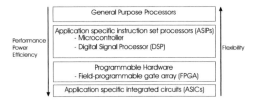

Fig. 3. Overview of different types of processors.

First of all there are the *General Purpose Processors* (GPPs). These processors are the most flexible ones and are highly optimized by a high amount of research. GPPs provide a high performance over a broad spectrum of applications and consume a lot of power. The high performance can be achieved by using parallel processing - techniques like pipelining and dynamic instruction scheduling are used - and multilevel caches. The main problem of these processors is that they cannot predict the run-time of processes caused by dynamic effects like scheduling, branch prediction and caching.

Application Specific Instruction-set Processors (ASIPs) are more application specific than general purpose processors. Advantages opposite GPPs are the lower costs and the lower power dissipation at a comparable performance.

Field Programmable Gate Arrays (FPGAs) and *Application Specific Integrated Circuits* (ASICs) are very application specific types of processors. They have only a small on-chip memory, a very low power consumption and are relatively cheap. These kind of processors reach the highest performance power trade-off followed by the worst flexibility.

In this work we take a closer look on low power processors. Low power processors are usually used for battery powered applications such as cellular phones, PDAs and mobile computers. The battery lifetime imposes the demands of the overall power consumption of the system.

In most low power applications the processor load varies over time. A cell phone consumes in standby mode only little power for listening to incoming calls. In contrast, during a call, when the processor is in active mode, it requires a high processing speed. Adjusting the variation on processor load, most of low power applications provide different power modes. Commonly used are an active mode where full performance is available and an idle mode where the processor is switched off, while peripherals are still on. In some cases, designers individually enable or disable peripherals. This functionality further decreases the power consumption.

Following processor types are low power processors:

- ASIPs
- FPGAs
- ASICs

Table 1 presents an example from each category. For mainstream production ASIPs and ASICs are preferred. Other application processors are mostly ARM processors. FPGAs are not mentioned here, because they are usually used for rapid prototyping or a small number of custom designs. The processors from Texas Instruments (MSP430F155) and Motorola (DSP56852) provide a very low active power consumption. The power consumption of the Intel processor

(PX270) is quite high, but can be reduced by using techniques like frequency and voltage scaling. Tensillica is situated between the DSP and the other ASIP. Power savings can obtained by the chip designer, thus it is important to spend sufficient effort in the chip design.

Category	MCU	DSP	other ASIP	ASIC
Processor	Texas Instruments MSP430F155	Motorola DSP56852	Intel PX270	Tensillica Xtensa LX
Clock frequency	8MHz	120MHz	312MHz	370MHz
Active power	10mW	122.4mW	374.5mW	246mW
Standby power	$3.5\mu W$	$36\mu W$	$1.722\mu W$	depends on design
Memory integration	16kBytes	20kBytes	322kBytes	determined by chip designer
Comments	includes HW MAC unit and DMA controller	includes DMA controller	supports speed/ voltage scaling	instruction set is customizable

Table 1. Overview of low power processors for signal-processing applications.

4 TEMPERATURE MODELLING

To overcome hot spots, temperature-aware design techniques are required. The idea of temperature-aware design is to consider the temperature during the entire design flow [3]. In former designs the temperature was only determined at the end of the flow. If the temperature was too high, a redesign of the chip was necessary. In new design flows it is possible to be aware of the temperature in each step. The University of Bologna has developed a thermal model for multi-processor system on chip (MP SoC) [1] to verify the die temperature. This model analyzes the thermal behavior of embedded systems, by considering multiple instances of cores and on-chip memory. Due to the fact that memories in SoCs are very important they are modeled at a finer granularity. The model also incorporates the package which provides a higher thermal resistance than high performance systems. Usually for SoC devices cheap packages with a bad thermal behavior are used. They are made for worst case temperature and most of them come without a heat spreader, so the heat has to be removed by natural convection. The goal of the thermal model is to simulate the temperature for given architectures. It is based on an equivalent RC-network, similar to HotSpot from the university of Virginia [9].

The thermal model is integrated in a complete and scalable multi-processor System-on-Chip platform called MPARM and provides cycle-accurate architectural simulation (see figure 4). Its purpose is the system-level analysis of design trade-offs in the usage of different processors, interconnects, memory hierarchies and other devices. MPARM output includes accurate profiling of system performance, execution traces and power estimation.

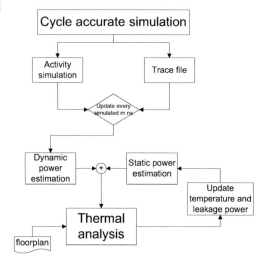

Fig. 4. Performance/energy/thermal simulation in a cycle-accurate SoC.

The platform allows to simulate the chip temperature at system level. The temperature influence on the performance can be analyzed with real applications. The big disadvantage of MPARM is the high granularity used for the simulation. So it takes a long time to gain results. In this work we have simplified this model to speed up the simulation. In a further step we have inserted a temperature sensor to measure the temperature of a certain block of the die at runtime. We have used a multimedia application to explore the impact of power and temperature-aware techniques. A simple application is a Finite Impulse Response (FIR) filter. A FIR filter with two sets of coefficients have been implemented.

5 DISCUSSION

We have worked on three experiments with different temperature policies. In a first step we illustrated that running the application on MPARM without a policy rises the temperature until thermal runaway or damage is caused. In a second step we presented the temperature behavior of the application when we are using frequency scaling. If the temperature exceeds a critical threshold we scale down the frequency. A reduction of the computational complexity is the result, but the duration of the computation takes a longer time. The point is that frequency scaling alone delivers not enough effort. Timing violations and signal distortions of the filter output may occur. So it is crucial to use a combination of Quality of Service (QoS) methods and frequency scaling. Our QoS method is to reduce the number of filter coefficients, if the temperature exceeds a critical threshold and also to scale down the frequency. A reduction of filter coefficients additionally reduces the computational complexity and the signal accuracy, but avoids signal distortions and timing violations. Figure 5 illustrates a comparison of the temperature by using the different policies.

Results are showing that temperature-aware design is more than simple frequency scaling, if we want to be sure that an application works correctly at real time.

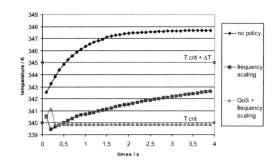

Fig. 5. Comparison of the temperature in the memory with no policy, frequency scaling and a combination of QoS and frequency scaling.

6 CONCLUSION

In this work we illustrated the importance of temperature-aware policies for the processor load. We started the work with studying conference papers of IEEE conferences to get an overview of existing low power and high performance architectures and memory cells. We specially took a look at their leakage power consumption and the technology node. Further we observed the sources of leakage currents and their reasons. A fundamental contributor will be the progressive trend in scaling transistors. Leakage currents are dependent on the temperature and will rise with increasing temperature. For future processor designs it is crucial to be aware of the temperature.

We implemented a simple multimedia application, a FIR filter, for a low power SoC. With this filter application we have explored the temperature variation across a die with and without using temperature-aware policies. Without any policy the temperature rises up steadily. This decreases the reliability of the system and can cause thermal damage. In a further step we explored the impact of frequency scaling on the system. Frequency scaling can cause timing violations and signal distortions. Finally, we investigated the influence of a combination of QoS method and frequency scaling. Our QoS method was to reduce the number of filter coefficients. This reduction decreases the computational complexity and influences positively the energy consumption of the processor. Energy consumption will decline and the processor will heat up much slower.

7 ACKNOWLEDGEMENT
This work was supported by Interuniversitary Microelectronics Center (IMEC) Leuven, Belgium and University of Applied Sciences in Hagenberg, Austria. Special thanks to Paul Marchal, Mario Huemer and Michael Bogner for their support during this work.

8 REFERENCES
[1] BENINI, L., BERTOZZI, D., BRUNI, D., DRAGO, N., FUMMI, F., AND PONCINO, M. Legacy systemc co-simulation of multi-processor systems-on-chip. In *Proc. of IEEE International Conference on Computer Design (ICCD)* (Fort Collins, CO, 2002).

[2] CHANDRAKASAN, A., BOWHILL, W., AND FOX, F. *Desing of High-Performance Microprocessor Circuits*. IEEE Press, New York, 2001.

[3] HUANG, W., STAN, M., SKADRON, K., SANKARANARAYANAN, K., GHOSH, S., AND VELUSAMY, S. Legacy systemc co-simulation of multi-processor systems-on-chip. In *Proc. of Design Automation Conference (DAC)* (San Diego, CA, 2004).

[4] INTERNATIONAL ROADMAP C. International technology roadmap for semiconductors. Tech. rep., Semiconductor Industry Association, 2003.

[5] KIM, N., FLAUTNER, K., BLAUUW, D., AND MUDGE, T. Circuit and mircoarchitectural techniques for reducing cache leakage power. *IEEE Transaction on Very Large Scale Integration (VLSI) Systems* (February 2004), 167–184.

[6] MAMIDIPAKA, M., KHOURI, K., DUTT, N., AND ABADIR, M. Leakage power estimation in sram's. Tech. rep., Department of Information and Computer Science, University of California, Irvine, CA, 2003.

[7] MANNE, S., KLAUSER, A., AND GRUNWALD, D. Pipeline gating speculation control for energy reduction. In *Association for Computing Machinery (ACM)* (Boulder, CO, 1998).

[8] ROY, K., MUKHOPADADHYAY, S., AND MAHMOODI-MEIMAND, H. Leakage current mechanims and leakage reduction techniques in deep submicrometer cmos circuits. *Proceedings of the IEEE* (February 2003), 305–327.

[9] SKADRON, K., STAN, M., HUANG, W., VELUSAMY, S., SANKARANARAYANAN, K., AND TARJAN, D. Temperature-aware micro-architecture. In *Proc. of IEEE 30th International Symposium of Computer Architecture (ISCA)* (San Diego, CA, 2003).

Index of Authors

Altendorfer Klaus, 139
Antely Angus, 66
Bäck Sabine, 83
Backfrieder Werner, 69, 74
Boido Claudio, 35
Brunner Uwe, 83
Dillinger Oliver, 188, 210
Ecker Jürgen, 188
Edlinger Günter, 66
Engelhardt-Nowitzki Corinna, 100, 115
Fasano Antonio, 35
Friedman Doron, 66
Fröber Ulrike, 57
Froschauer Stefan, 60
Gaupmann Andreas, 204
Gierlinger Sebastian, 188
Greisberger Wolfgang, 100
Griebenow Sirid, 53
Guger Christoph, 66
Hable Oliver, 175
Hainisch Reinhard, 60
Hauer Barbara, 175
Heiligenbrunner Martina, 194
Horwath-Winter Jutta, 53
Jodlbauer Herbert, 139, 150
Kleebauer Peter, 188
Kraigher-Krainer Jörg, 3, 22
Kronberger Gabriel, 153, 160
Kurschl Werner, 216
Lackner Elisabeth, 115
Langer Josef, 188, 210
Leeb Robert, 66
Liebmann Hans-Peter, 22
Lutherdt Stefan, 57
Madlmayr Gerald, 188, 210
Marek József, 95
Maschek Wilma, 69
Mitsch Stefan, 216
Müller Alexander, 57
Németh Péter, 95
Pfeifer Franz, 74
Plachy Thomas, 224
Prokop Rene, 116
Promberger Wolfgang, 150
Pucher Roland, 188
Rieger Gebhard, 53
Rohrhofer Evelyn, 127
Schaffer Christoph, 188, 210
Schmitz Klaus, 89
Schmut Otto, 53

Schöffer Martin, 166

Schöffl Harald, 60

Schönböck Johannes, 116

Slamanig Daniel, 194

Slater Mel, 66

Spaeth Karin, 127

Stingl Christian, 194

Stüger Alexander, 107

Swoboda Roland, 74

Vinayagamoorthy Vinoba, 66

Vymazahl Dieter, 188

Wallinger Johann, 180

Wallner Tatjana, 107

Weger Christian, 166

Weidenhiller Andreas, 153, 160

Witte Hartmut, 57

Zauner Martin, 60

Zsifkovits Helmut E., 115

Zwettler Gerald, 74